Advances in Lectin Research   Volume 2

Advances in Lectin Research

Volume 2

Edited by Hartmut Franz

Coeditors: Ken-ichi Kasai, Jan Kocourek,
Sjur Olsnes, Leland M. Shannon

With 35 Figures

Springer-Verlag Berlin Heidelberg GmbH

Hartmut Franz
Staatliches Institut für Immunpräparate und Nährmedien
Klement-Gottwald-Allee 317-321, Berlin, DDR-1120

Ken-ichi Kasai
Faculty of Pharmaceutical Sciences, Teikyo University
Sagamiko, Kanagawa 199-01, Japan

Jan Kocourek
Dept. of Biochemistry Charles University Prague
Hrusicka 2515, 14100 Praha 4 – Spořilov 11, Czechoslovakia

Sjur Olsnes
Norsk Hydro's Institute for Cancer Research, Montebelle Oslo 3, Norway

Leland M. Shannon
University of California, Riverside, California 92521, USA

Sole distribution rights for all non-socialist countries
Springer-Verlag Berlin Heidelberg New York London Paris Tokyo

ISBN 978-3-662-11062-1   ISBN 978-3-662-11060-7 (eBook)
DOI 10.1007/978-3-662-11060-7

Typesetting, printing and binding: Grafische Werke Zwickau
2131/3140-543210

Authors

Prof. Dr. Carl A. K. Borrebaeck
Department of Biotechnology, University of Lund, P. O. Box 124
S-22100 Lund, Sweden

Dr. Roland Carlsson
Bioinvent International AB
S-22370 Lund, Sweden

Dr. Yaeta Endo
Department of Biochemistry, Yamanashi Medical College, Tamaho
Nakakoma, Yamanashi 409-38, Japan

Prof. Dr. Dr. Hartmut Franz
Staatliches Institut für Immunpräparate und Nährmedien
Klement-Gottwald-Allee 317-321, Berlin, DDR-1120

Dr. Hans-Joachim Gabius
Max-Planck-Institut für experimentelle Medizin, Abt. Chemie
Hermann-Rein-Straße 3, Göttingen, D-3400

Dr. Arpad Pusztai
Rowett Research Institute, Buckeburn
Aberdeen, AB2 9SB, Great Britain

From the Foreword to Volume 1

The years 1987/88 appear to be a suitable moment for the publication of "Advances in Lectin Research". The detection of ricin 100 years ago led to interesting results concerning sugar-binding proteins, now called lectins. Today two main lines exist in "lectinology". Firstly, lectins are used as multi-purpose tools (analysis and isolation of glycoconjugates, characterization and preparation of cells and microorganism, lectin histochemistry). Secondly, lectins are of interest concerning their physiological roles and other biological activities. Especially during the last 10 years, the latter aspects have been widely noticed. More and more the opinion is being accepted that lectins are glues of moderate affinity, making possible the contact of biologically cooperating glycoconjugates and cells bearing glycoconjugates.

The results of lectin research anticipated for the coming years should be of great importance not only for basic research, but also for diagnostic and even therapeutic application (adherence inhibition of bacteria and tumor cells, immunomodulation, lectin-mediated cytotoxicity, chemically modified toxic lectins) ...

"Advances in Lectin Research", as a collection of review articles, is not in competition with textbooks and congress proceedings ...

Foreword to Volume 2

The first Volume of "Advances in Lectin Research" met kind reception. This has encouraged us to present the second Volume one year later dealing with such different aspects as mitogenicity of lectins, Viscaceae lectins, the mechanism of the inactivation of eukaryotic ribosomes by toxic lectins, nutritional effects of lectins and effects of tumor lectins.

Illustrations of lectin producing plants will be continued in Volume 3.

It is again the duty of the editor to thank all authors, the coeditors and the lector of the "Verlag Volk und Gesundheit", Mrs. Margitta Hintz for their support. The help of my colleague Dr. Rolf Wachowius was important for me. I thank Dr. Michael Gelbin, Dr. Peter Ziska and Heinz Zorn for reviewing the manuscript. I am indebted to Mrs. Marianne Lobstein for her help in the final preparation of this manuscript.

Hartmut Franz, Berlin

Contents

1 Lectins as Mitogens

Carl A. K. Borrebaeck and Roland Carlsson

1.1 Introduction

The cell surface of mammalian lymphocytes contains several hundreds of membrane-bound proteins, many of which are glycosylated. Very few have been identified and characterized and the function of only a handful of these cell surface proteins is known. Carbohydrate binding proteins, e.g. lectins, have thus a large number of different surface-bound cell ligands with which to interact. The most dramatic consequence, resulting from the interaction of some lectins with membrane-bound glycoproteins, is the polyclonal activation of small resting lymphocytes to rapidly dividing cells. This mitogenic effect can be broken down into a number of complex cellular events, some of which will be discussed below.

1.2 Mitogenic Lectins

1.2.1 General Features

The first lectin that was reported to be mitogenic was phytohemagglutinin (PHA), the seed lectin isolated from red kidney beans, *Phaseolus vulgaris* (Nowell 1960). During the following decade three additional mitogenic lectins were found. These were pokeweed mitogen isolated from *Phytolacca americana* (Farnes et al. 1964), *Wistaria floribunda* mitogen (Barker and Farnes 1967) and concanavalin A from *Canavalia ensiformis* (Douglas et al. 1969). Presently, over 30 different mitogenic plant lectins have been reported (Table 1.1) and there are good reasons to believe that all lectins are mitogenic if tested under the proper conditions (Glad and Borrebaeck 1984; S. A. Möller et al. 1986).

The isolectin of PHA, containing four E (E = "erythroreactive") subunits (PHA-E_4), was for example long considered to be a non-mitogen for lymphocytes but a strong hemagglutinin of erythrocytes. However, PHA-E_4 possesses a strong mitogenic potency if all serum glycoproteins present during cell culture are removed (Glad and Borrebaeck 1984). It was shown that PHA-E_4, which has a broader carbohydrate specificity than the -L_4 isolectin, had an affinity for 14 identified human serum glycoproteins, whereas PHA-L_4 interacted with only 9 of these glycoproteins (Glad and Borrebaeck 1984). If PHA-E_4 was subsequently tested in serum devoid of glycoconjugate it showed a strong mitogenic effect with a dose-response maximum around 1–2 $\mu g \cdot ml^{-1}$. The mitogenic profiles of the two homologous PHA isolectins tested in medium containing 10% fetal calf serum or under serum-free conditions are shown in Figure 1.1. The effect of serum glycoconjugate has also been shown to have an effect on the mitogenic potency of various *Lathyrus* lectins (Borrebaeck and Rougé 1986), where the presence of serum glycoconjugate reduced the mitogenic potency of the lectins five to ten fold.

Table 1.1 Mitogenic plant lectins

Lectins	Cell specificity	References
Abrus precatorius	n. d.*	Closs et al. 1975
Agaricus campestria	n. d.	Presant and Kornfeldt 1972
Bauhinia carronii	n. d.	Curtain and Simons 1972
Concanavalin A	T/B	Novogrodsky and Katachalski 1973, Möller et al. 1986
Hura crepitans	T	Falasca et al. 1980
Lathyrus aphaca	n. d.	Borrebaeck and Rougé 1986
articulatus	n. d.	Borrebaeck and Rougé 1986
cicera	n. d.	Borrebaeck and Rougé 1986
ochrus (whole lectin)	n. d.	Borrebaeck and Rougé 1986
ochrus I	n. d.	Borrebaeck and Rougé 1986
ochrus II	n. d.	Borrebaeck and Rougé 1986
hirsutus	n. d.	Borrebaeck and Rougé 1986
odoratus	n. d.	Kolberg 1978
sativus	n. d.	Gupta et al. 1980
tingitanus	n. d.	Borrebaeck and Rougé 1986
vernus	n. d.	Borrebaeck and Rougé 1986
Lentil	T/B	Miller 1983, Toyoshima et al. 1970
Lima bean	T/(B)	Besseler et al. 1976, Ruddon et al. 1974
Maclura pomifera	n. d.	Horejsi and Kocourek 1978
Pea	n. d.	Trowbridge 1973
Peanut agglutinin	n. d.	Novogrodsky et al. 1975
PHA-L$_4$	T	Felsted et al. 1975
PHA-E$_4$	T/(B)	Glad and Borrebaeck 1984, Möller et al. 1986
Phaseolus coccineus	n. d.	Angelisova and Haskovec 1978
Phaseolus vulgaris (PHA)	T	Nowell 1960
Pokeweed mitogen	T/B	Farnes et al. 1964
Rice bran	n. d.	Tsuda 1979
Robinia pseudoaccacia	T/B	Hořejší and Kocourek 1978, Schumann et al. 1973
Sarothamnus scoparius	n. d.	Hořejší and Kocourek 1978
Sophora japonica	n. d.	Terao and Osawa 1973
Soybean agglutinin	T	Novogrodsky and Katachalski 1973
Ulex europeus I	T	Schumann et al. 1973
Ulex europeus	B	Yamaguchi et al. 1981
Vicia faba	n. d.	Wang et al. 1978
Vicia sativa	n. d.	Falasca et al. 1979
Wax bean	n. d.	Douglas et al. 1969, Sela et al. 1973
Wheat germ agglutinin	T/B	Brown et al. 1976, Greene et al. 1981c, Udey et al. 1980
Wistaria floribunda	n. d.	Toyoshima et al. 1971

*: not determined

The majority of the lectins listed in Table 1.1 only activate the T cell lineage and are not directly mitogenic for B lymphocytes. A few plant lectins that stimulate both B and T cells are also known, e.g. Pa-I isolectin of pokeweed mitogen (Basham and Waxdal 1975), lentil lectin (Mil-

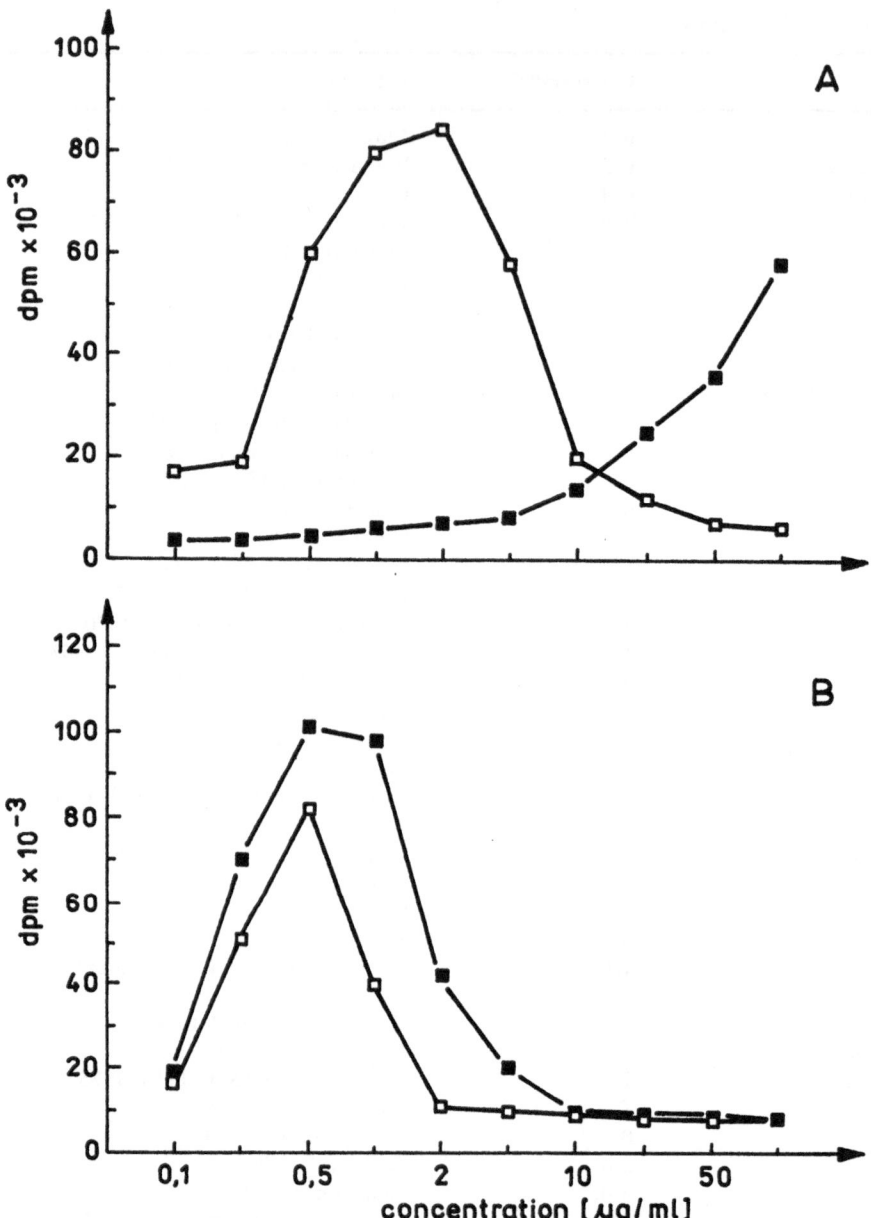

Fig. 1.1 PHA-E4 (A) and -L4 (B) induced (methyl-^{3}H)-thymidine incorporation of human peripheral lymphocytes cultured in RPMI 1640 medium, supplemented with 10 % fetal calf serum (■) or under serum-free conditions in HL-1 medium containing no serum supplement (□)

dpm: decay per minute

ler 1983) and wheat germ agglutinin (Udey et al. 1980; Greene et al. 1981c; Ulmer et al. 1982). However, the lectins from the crab *Homarus americanus* (Campbell et al. 1982) and from the slime mold *Dictyostelium purpureum* (Lipsick et al. 1980) are examples of non-plant lectins that stimulate mouse B cells but not T cells.

1.2.2   Structurally Related Mitogenicity

The cellular response to a lectin is affected by enzymes, e.g. glycosidases. Soybean agglutinin is non-mitogenic to mouse lymphocytes unless the cell surface is treated with sialidase (Novogrodsky and Katchalski 1973). In addition the mitogenic properties of lectins can be effected by chemical modifications, e.g. changing the protein structure and/or the valency of the lectin. Lima bean lectin exists as two isolectins, the tetravalent isolectin (MW 247,000) is a strong mitogen for human and bovine lymphocytes. On the other hand, the divalent isolectin (MW 124,000) exhibits only a very weak mitogenic potency (Ruddon et al. 1974; Bessler et al. 1976; Munske et al. 1981; Pandolfino et al. 1983). Similarly, the tetrameric soybean agglutinin ist not mitogenic for lymphocytes from a number of different species; sialidase treatment does not improve the mitogenic potency (Schechter et al. 1976). Polymerization of the native lectin will, however, convert it into a mitogen for untreated pig lymphocytes and for sialidase-treated human, mouse and rat lymphocytes (Lotan et al. 1973, 1975). Contrary to these findings that the mitogenic potency is related to the multivalency of the lectin tetrameric Con A, it is not a more potent mitogen as compared to divalent Con A. Even monomeric Con A, prepared by a combination of succinylation and photoaffinity labeling has been reported to be mitogenic (Beppu et al. 1975; Fraser et al. 1976; Wang and Edelman 1978; Saito et al. 1983) although it has been argued that this mitogenic effect might be due to residual amounts of oligomeric Con A (Lis and Sharon 1986). The involvement of the hydrophobic sites of Con A (Edelman and Wang 1978) has not been studied but might be involved in a multivalent cellular interaction. Interestingly, when Ledbetter et al. (1986) and Tsoukos et al. (1985) studied the effect of multivalency on the mitogenic potency of monoclonal antibodies directed against a cell surface glycoprotein (T3) known to be involved in T cell activation (see below), similar results were obtained, as described above for Con A. Multivalency was obtained by immobilizing anti-T3 antibodies onto Sepharose. These immobilized antibodies caused resting T cells to express high levels of functional IL-2 receptors. Divalent F(ab')$_2$ and monovalent Fab fragments also enhanced T cell activation but required additional signals to fully activate the T cells (Tsoukos et al. 1985; Ledbetter et al. 1986). Interestingly, the F(ab')$_2$ and Fab fragments caused a rapid down-regulation of the T3 expression whereas immobilized anti-T3 antibodies did not. Cells with down-regulated T3 became unresponsive to further challenge with mitogenic lectins and antibodies.

Recently, the mitogenic properties of a group of phylogenetically closely related lectins belonging to the tribe Vicieae were investigated (Borrebaeck and Rougé 1986). These lectins from the genus *Lathyrus* all possess an $\alpha_2\beta_2$ structure and have been shown to contain striking similarities in physicochemical and biological properties. This was subsequently corroborated by extensive homology among the amino acid sequences of both their light and heavy subunits (Richardson et al. 1984; Yarwood et al. 1985). The lectins have, furthermore, very similar carbohydrate specificities and the two *Lathyrus ochrus* isolectins I and II exhibit identical specificities (Richardson et al. 1984). Interestingly, the different *Lathyrus* lectins showed a marked difference in their abilities to induce lymphocyte transformation of isolated human peripheral lymphocytes (Borrebaeck and Rougé 1986). The optimal mitogenic dose of the different lectins differed by a factor of 10–30. There was also a difference in mitogenic properties of *L. ochrus* (whole lectin) and the isolated isolectins *L. ochrus* I and II. The whole lectin had a considerably (3–10 times) better ability to induce cell growth than the individual isolectins. It seemed that the mixture of the two isolectins, which is represented by the whole lectin, had a synergistic effect on lymphoid cells (Borrebaeck and Rougé 1986). This difference in mitogenic properties had no obvious explanation.

13

The slight differences observed among the amino acid sequences of constituent light and heavy subunits (only in a few positions) of the different *Lathyrus* lectins, most probably induce small conformational changes. This could consequently lead to differences in their affinities toward complex cell surface-bound glycoconjugates (Rougé et al. 1987). Differences in carbohydrate-binding ability was also reported for the *Dolichos biflorus* seed lectin, which consists of equal amounts of two subunit types (subunit I: MW 27,700, and II: MW 27,300) (Etzler et al. 1986). Subunit I had the ability to bind N-acetylgalactosamine residues, whereas subunit II dit not; only ten amino acids differed in their carboxyl terminal ends (Etzler et al. 1981). Furthermore, the comparison of the amino acid sequences of the *Dolichos biflorus* seed lectin and the β-sub-unit of DB58 (*Dolichos biflorus* 58, formerly called CRM) (Etzler et al. 1986) indicated that these two similar lectins originated from separate genes and showed a 85% homology (E. Etzler personal communication). The most striking result, with regard to the structure-function relation of mitogens, is that DB58 is a mitogen, whereas the seed lectin has no mitogenic properties (Borrebaeck and Etzler 1982).

1.3   Lectin Induced Mitogenicity

The polyclonal activation induced by a lectin mitogen can commit up to 70–80% of a certain lymphocyte population (Hume and Weidemann 1980a). The first step in the induction of mitogenicity is binding of the lectin to cell surface glycoconjugates. This is a necessary but not a sufficient step since several non-mitogenic lectins exhibit a binding to lymphocytes without the ability to induce mitosis (Dillner-Centerlind et al. 1980). The triggering of quiescent, non-dividing lymphocytes to actively growing cells can be divided into events like (1) an induction/activation phase, driving the cells from G0 to early G1 in the cell cycle, (2) a proliferation phase, in which activated cells go through a S-phase with DNA synthesis, (3) a differentiation phase, during which T and B cells acquire specialized abilities, e.g. cytotoxic functions and antibody production.

1.3.1   Lectin-Membrane Interactions

Since lectins recognize many glycosylated cell surface receptors it is not known with certainty which of these molecule(s) conveys the activation signal. The mitogenic lectin *per se* does not stimulate the lymphocyte to proliferate, but induces the production of cellularly derived growth factors and the expression of receptors for these factors. A summary of the cellular events resulting from the lectin-receptor interaction is given in Figure 1.2.

The molecular interactions leading to activation and proliferation of T lymphocytes after antigenic stimulation are today fairly well understood. Antigens in combination with cell surface molecules encoded by the major histocompatibility complex, i.e. class I and II antigens, are recognized by the T cell receptor (Ti). An activation signal is then transmitted to a molecular complex called CD3 (T3), consisting of three polypeptide chains. This complex is embedded in the membrane close to but not covalently associated with the T cell antigen receptor (Alcover et al. 1987). The signal is then conveyed to phospholipase C which converts phosphatidylinositol biphosphate to inositoltriphosphate (IP$_3$) and diacylglycerol (DAG). These two products are potential activators of protein kinase C which ultimately results in the activation of the cell (for review see Isakov et al. 1986). Activation of T cells can also be achieved by using monoclonal antibodies directed against the Ti-T3 molecular complex, either recognizing the T cell receptor or

14

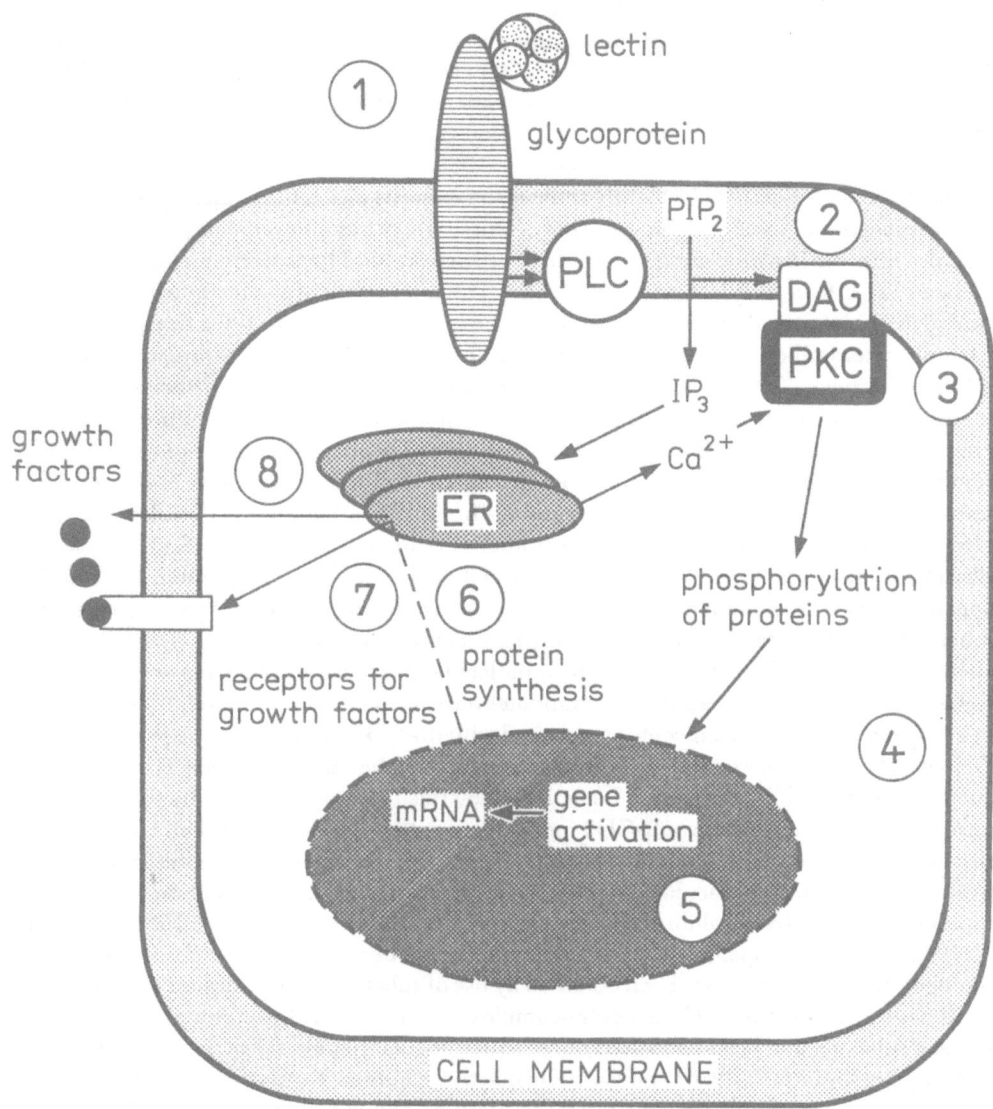

Fig. 1.2 Schematic representation of lectin-induced activation of lymphocytes

In a first step the lectin interacts with surface-bound glycoproteins (1). This leads to activation of phospholipase C (PLC), which hydrolyzes phosphatidylino-sitolbiphosphate (PIP_2) into diacylglycerol (DAG) and inositoltriphosphate (IP_3) (2). IP_3 then mobilizes Ca^{2+} from intracellular sources, e.g. endoplasmic reticulum (ER). Ca^{2+} together with DAG activates (PKC) (3), which phosphorylates a number of proteins (4). This leads to activation of specific genes and the production of mRNAs (5), resulting in protein synthesis (6). This includes receptors for growth factors, such as IL-2; several of these receptors are integral membrane proteins (7). Growth factors are also synthesized and secreted (8).

the CD3 complex (Van Wauwe et al. 1980; Spits et al. 1985). Recently it was also demonstrated that T cells could be activated via an alternative pathway utilizing monoclonal antibodies against a pan-T cell surface glycoprotein of 50,000 MW called CD2 (Meuer et al. 1984). The signal transduction system engaged in this activation process shows similarities but also dissimilarities to the one used by T cells after activation via the Ti-T3 complex, and possibly represents a distinct pathway for activation (June et al. 1986). Mitogenic lectins have been used in T cell activation as a means to polyclonally activate the majority of T cells (Hume and Weidemann 1980), irrespective of their antigen specificity. This strategy has allowed analysis of many processes also involved in clonal antigen-specific T cell activation. The most frequently used lectins have been PHA, Con A, PWM and lentil lectin. The interaction of the lectin with membrane bound glycoconjugates on the T cell leads to, in the presence of monocytes, a transition of resting lymphocytes in the G0 phase to enlarged blastoid cells in G1 phase, expressing a panel of new surface antigens. These antigens are specific for the activated T cells. Furthermore, the activated T cells produce an array of lymphokines that interact in numerous immunological cellular networks. To further drive the activated T cell from G1 phase into the S phase of the cell cycle the interaction of T cell growth factors (IL-2, IL-4) with their receptors is a prerequisite (Cantrell and K. A. Smith 1984; Mosmann and Coffman 1987). This interaction also modulates further lymphokine production e.g. synthesis and secretion of interferon (IFN-γ) (Vilcek et al. 1985; Dohlsten et al. 1986). The majority of cell surface glycoproteins interacting with a mitogenic lectin is most probably irrelevant for the activation of the T cell. As a consequence, it has been difficult to identify the lectin receptors directly involved in the activation process. Hitherto no definitive demonstration of these molecules has been reported. However, using indirect approaches two or three possible candidates have been identified. As mentioned above, T cells recognize antigens via a receptor associated with the CD3 complex. The expression of this Ti-T3 complex is down-regulated by monoclonal antibodies against either Ti or T3. Consequently, modulated T cells cannot be activated by this pathway as shown by the lack of response to antigen or mitogenic anti-CD3 monoclonal antibodies (Kammer et al. 1984). T cells, with a down-regulated Ti-T3 complex, have been used to probe whether some mitogenic lectins activate T cells via this molecular complex. It has been shown that modulation of the Ti-T3 complex greatly impairs the ability of T cells to respond to PHA (Kammer et al. 1984; Ceuppens et al. 1986), PWM (Ceuppens et al. 1986) or Con A (Ledbetter et al. 1986; Carlsson et al. 1987b submitted). Similar results have been obtained by use of subclones of the human T cell line Jurkat. Clones expressing the CD3 molecular complex, produce IL-2 after stimulation with PHA, whereas a subclone that lacks CD3 does not produce this interleukin (Weiss et al. 1984). These results suggest that PHA, PWM and Con A can activate human T cells via the Ti-T3 complex. Interestingly, it was recently demonstrated that PHA-P (which contains mostly PHA-L$_4$) activates T cells via CD2, whereas PHA-M (which contains mostly PHA-E$_4$) activates T cells via CD3 (O'Flynn et al. 1986). Different glycosylation patterns of CD2 and CD3 matching the carbohydrate specificities of PHA isolectins have, however, not been demonstrated. Furthermore, PHA-P and -M were not demonstrated to be pure homologous isolectins like PHA-L$_4$ and -E$_4$. PHA, WGA and other lectins (see below) have been reported to activate T cells via carbohydrate moieties on the antigen receptor (Ti) (Chilson et al. 1984), and biochemical data showing physical interaction between PHA and the receptor was also evident. However, caution should be observed when interpreting experiments with modulation of Ti-T3 since down-regulation of this complex impairs activation signals, not only via Ti-T3 but also via the unrelated CD2, Tp44 (Ledbetter et al. 1986) and Tp 103 cell surface molecules (Fleischer 1987). Thus, down-regulation of the Ti-T3 complex also induces a cellular state of unresponsiveness to further activation via other surface antigens. The reason for this is presently unknown but it has been suggested

16

that the activation pathways utilize a common pathway for activation (Yang et al. 1986). However, it has been clearly shown that T cell variants lacking the CD2 molecule can still be activated by anti-CD3 antibodies and vice versa (Moretta et al. 1987). Furthermore, some lectins have been suggested to activate human T cells via still other cell-surface antigens (Ceuppens et al. 1986; Yachie et al. 1987). The evidence that PWM can utilize a pathway other than Ti-T3 is based on experiments with modulation of this complex. Modulated cells respond well to high doses of PWM but not to low doses of the lectin, suggesting the presence of one Ti-T3 dependent and one independent pathway. Interestingly, activation with high doses of PWM is apparently IL-2 independent since no IL-2 production can be detected and monoclonal antibodies against the IL-2 receptor cannot block the PWM induced proliferation. It is possible that these T cells are induced to mainly produce and utilize a recently described multipotent growth factor, interleukin-4 (IL-4) (Mosmann and Coffman 1987). WGA has also been reported to activate human T cells via cell-surface molecules distinct from the Ti-T3 complex (Yachie et al. 1987). These conclusions are based on tests with not only Ti-T3 modulated T cells but also with T cell mutants lacking expression of the Ti-T3 complex.

Thus, it seems clear that the binding of lectins to monomorphic determinants of the Ti-T3 complex will bypass an antigen-specific activation and activate the T cells polyclonally instead. It is furthermore obvious that some lectins exert their mitogenic activity through other surface molecules; whether these molecules serve any function in antigenic activation awaits further investigations.

1.3.2   Lectin Induced Inhibition of Cell Proliferation

Apart from being lymphocytic mitogens a few lectins have also been reported to act anti-mitogenically, i.e. preventing blast formation of normal lymphocytes in the presence of antigens or other lectins. *Agaricus bisporus* (Greene et al. 1981b), tomato lectin (Kilpatrick et al. 1986) and WGA (Greene and Waldmann 1980) have all been suggested to act as anti-mitogens. The mechanisms of anti-mitosis are still mostly unknown, although Greene and co-workers suggested the existence of "suppressor receptors" in human lymphocytes (Greene et al. 1981b). The existence of receptors which mediate the transmembrane signalling that down-regulates cellular proliferation, implies that an active, energy-dependent series of cellular events take place. Recently this was reported to be the case for several human leukemic cell lines (Borrebaeck and Schön 1987), where anti-proliferative receptors were partially characterized. The anti-mitotic or mitotic interaction of WGA with human lymphocytes has been the subject of conflicting reports for a decade. WGA has been found to be a non-mitogen (Boldt et al. 1975), a mitogen for B cells (Greene et al. 1981c), a mitogen for a subset of T cells (Van Wauwe et al. 1980), an anti-mitogen (Greene et al. 1976, 1981b) and a combination of all these activities (Kilpatrick and McCurrach 1987). The anti-mitotic effects of WGA have been suggested to be the consequence of activated T suppressor cells (see below), which could also explain the "stimulation-inhibition paradox" (McClain and Edelman 1976; Hume and Weidemann 1980). Kilpatrick and McCurrach suggested, however, that the contradictory results found in the literature were due to experimental variations (Kilpatrick and McCurrach 1987) since the mitogenicity of WGA was dependent on
— lymphocyte donors
— presence of platelets
— serum source for in vitro cell culture, and
— presence of accessory cells (MØ).

17

The anti-mitotic effect was also suggested to be caused by the binding of WGA to IL-2 receptors (Reed et al. 1985; Kilpatrick and McCurrach 1987) and not by suppressor cells. In summary, it seems plausible that earlier contradictions of the effect of WGA were due to experimental artifacts and that the cause of the anti-mitotic effect still needs to be clarified.

PHA-L$_4$ was recently shown to exert a strong anti-proliferative effect on human acute and chronic lymphocytic leukemic cell lines of the T cell type (T-ALL/CLL) (Borrebaeck and Schön 1987). Proliferation and DNA synthesis were inhibited in a dose-dependent fashion and PHA-L$_4$ induced a significant anti-proliferative response of CCRM-CEM and MOLT-4 tumor cells at a concentration of as low as 0.05 µg · ml^{-1}. A 50% inhibition of DNA synthesis of these two cell lines was obtained at 0.4 and 0.5 µg PHA-L$_4$ · ml^{-1}, respectively. The effect was cytostatic rather than cytotoxic. Measurements of the overall cellular metabolism as well as of glycolytic lactate and oxygen consumption during lectin treatment supported the fact that PHA-L$_4$-induced anti-proliferation was specific and energy dependent. The membrane components involved in this response were identified and partially characterized; PHA-L$_4$ did not bind the receptor for IL-2 (MW 55,000, Tac antigen) or transferrin (MW 85,000–90,000). In contrast, PHA-E$_4$ could not induce a specific anti-proliferative response of the same leukemic cell lines (Borrebaeck and Schön 1987), suggesting that the carbohydrate specificity of the two isolectins differs for complexed membrane bound glycoconjugates. These findings strongly support the existence of anti-proliferative-associated antigens and argues against the activation of T suppressor cells. Impaired ability of tumor cells to respond to anti-proliferative signals or the inability of other cells to give these signals might be involved in the etiology of some tumor diseases.

1.3.3   Lectins and Lymphocytes

1.3.3.1   Role of Lymphokines

Activation of T cells is at least a two-step event. In a first step, interaction of the lectin with the T cell leads to expression of IL-2 receptors. This occurs as the cells are driven from G0 to G1 phase of the cell cycle and is independent of accessory cells (Cantrell and K. A. Smith 1984; Davis and Lipsky 1985, 1986; Isakov et al. 1986). The receptor for IL-2 is induced in the major types of T lymphocytes, i.e. the CD4$^+$ T helper/-inducer and CD8$^+$ T cytotoxic/suppressor cells. The T cells produce IL-2 in response to at least one additional signal, i.e. interleukin-1 (IL-1) from an accessory cell (Davis and Lipsky 1985, 1986; Truneh et al. 1986). Interaction of IL-1 and the accessory cell with the lectin stimulated T cells results in production of IL-2 mainly from the T helper subset (Reinherz et al. 1980; Dohlsten et al. 1986). Interleukin-2 has multiple effects on lymphoid cells. It is a growth factor and enhances the activity of T cells, NK cells and B cells (Abo et al. 1983; Cantrell and K. A. Smith 1984; Itoh et al. 1985; Kishi et al. 1985; Vasquerz et al. 1986). IL-2 also modulates the production of other lymphokines, e.g. it stimulates production of interferon-γ (IFN-γ) by both CD4$^+$ and CD8$^+$ T lymphocytes (Vilcek et al. 1985; Dohlsten et al. 1986). The expression of interleukin-2 receptors is also enhanced by IL-2, although this primarily yields low-affinity receptors (K. A. Smith and Cantrell 1985). These receptors cannot transduce the growth-promoting activity of IL-2 and they have been suggested to exhibit a negative regulatory function on T cell growth (Kumar et al. 1987). Activation of lymphocytes produces other lymphokines like interleukin-3, interleukin-4 [also called B cell growth factor-I (BCGF-I) or B cell stimulatory factor-(BSF-1)] and interleukin-5 [also called T cell replacing factor (TRF) or B cell differentiation factor (BCDF)] (Mosmann and Coffman 1987). The ef-

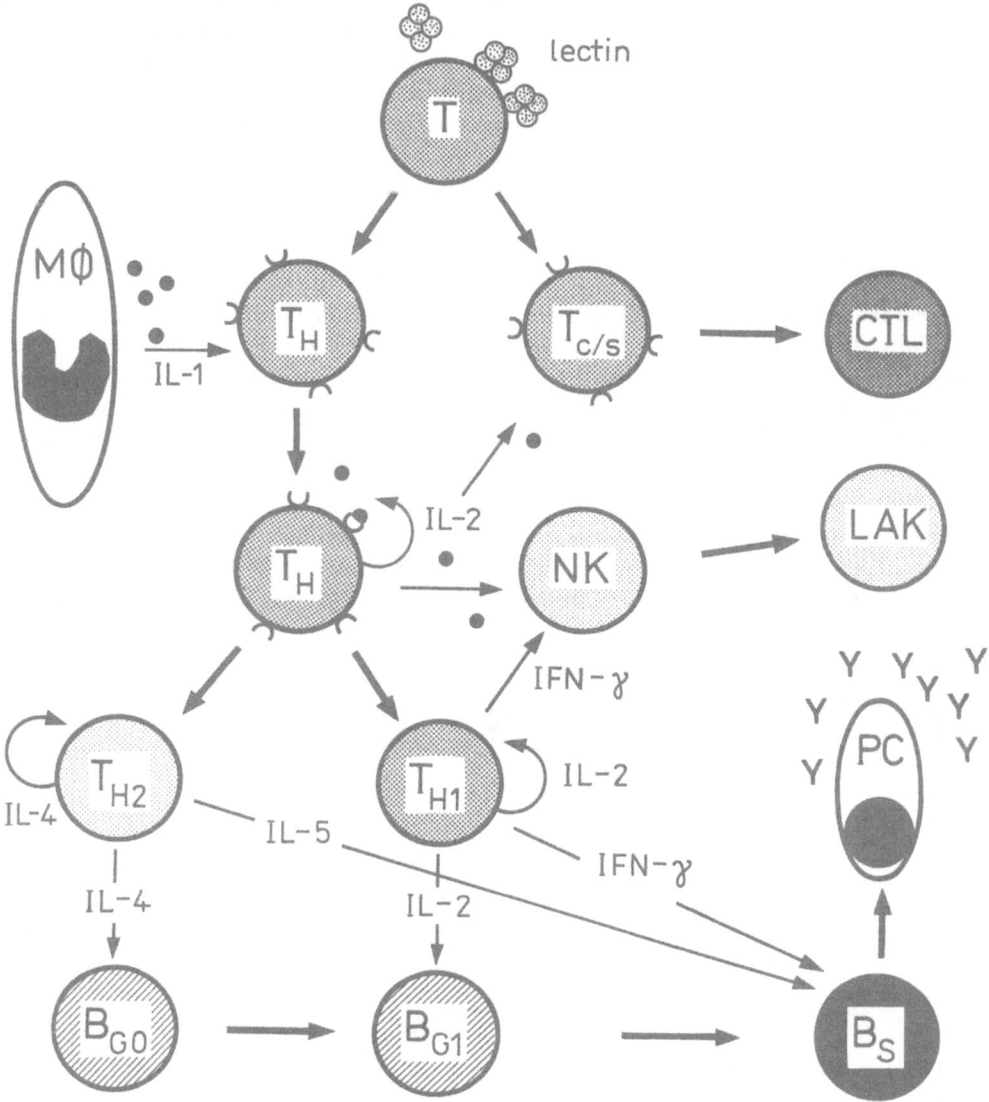

Fig. 1.3 Lectin-induced lymphokine production

Lectin interacts with the T cell independently of accessory cells, which leads to expression of IL-2 receptors on both T helper (T_H) and T cytotoxic/suppressor ($T_{C/S}$) cells. The T_H cells produce increased levels of IL-2 after stimulus from accessory cell-derived IL-1. IL-2 is an autocrine growth factor and also participates in the proliferation and differentiation of cytolytic T cells. Furthermore, IL-2 activates natural killer cells (NK cells) to cytolytic lymphokine-activated killer cells (LAK cells). T_H cells can, in the mouse, be subdivided in to T_{H1} and T_{H2}. These cells differ in their capacity to produce and utilize lymphokines. T_{H1} cells produce IL-2 and IFN-γ and use IL-2 as a growth factor, whereas T_{H2} cells produce IL-4 and IL-5; they use IL-4 as an autocrine growth factor. IL-4 also activates resting B cells to enter the cell cycle, where they become responsive to IL-2. This combination of IL-2 and IL-4 drives the B cells into S phase and proliferation. Finally, the B cells differentiate into antibody-producing plasma cells due, in part, to IL-5 and IFN-γ.

MØ: macrophage

fects of these factors were considered earlier to be quite discrete acting at specific stages during the proliferation and differentiation of B or T cells. It is, however, becoming increasingly clear that the action of these lymphokines are pleiotropic; they stimulate different cell types e. g. IL-4, which besides stimulating B cells also acts as an autocrine growth factor for some T cells and has effects on a variety of other cell types (Mosmann and Coffman 1987). Figure 1.3 summarizes the different effects of some lymphokines.

1.3.3.2 Con A as a B cell Mitogen

Con A was thought of as a specific T cell mitogen for a long period (Novogrodsky and Katchalski 1973). Hawrylowicz and Klaus (1984) showed, however, that Con A can activate/induce B cells but cannot induce DNA synthesis and cell divisions. At Con A doses which were optimal for a T cell response, the B cells depolarized, enlarged and displayed increased levels of membrane bound Ia antigens. Their data indicated that Con A causes purified, resting B cells to leave G0 and to enter G1 phase. The B cells failed, however, to induce DNA synthesis (Hawrylowicz and Klaus 1984). Recently, Con A was shown to be a strong mitogen for purified resting B cells if supported by allogenically derived helper factors (AHF) (S. A. Möller et al. 1986). The presence of AHF was shown to be necessary and sufficient during initiation of the cell culture to drive the Con A activated B cells into S phase and mitosis. The dose-response maximum could be ob-

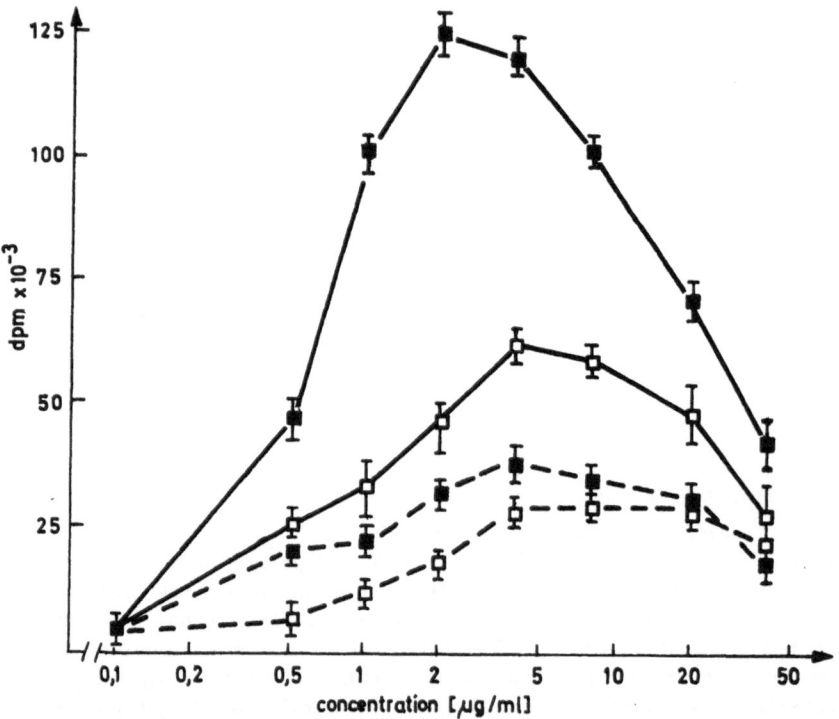

Fig. 1.4 Proliferation assay of CD4[+] (———) and CD8[+] (- - - -) cells, in supplemented RPMI 1640 medium containing 50 µM mercaptoethanol and 0.4% human serum albumin, PHA-L$_4$ (■) and PHA-E$_4$ (□)

dpm: decay per minute

20

tained after a 4-day culture using 25 % of a supernatant from a mixed lymphocyte culture as a source of AHF. The activation of optimal concentrations of Con A (2 μg · ml^{-1}) and AHF (25 %) could be completely inhibited by α-D-mannopyranoside (S. A. Möller et al. 1986). The Con A dependent B cell activation might be due to cross-linking of surface IgM molecules, eventually making the cells responsive to T or B cell derived helper factors. The nature of these helper factors is as yet not fully characterized (Danielsson et al. 1986).

1.3.3.3 Interaction of PHA Isolectins with Subsets of Human T Cells

The effect of PHA-E$_4$ and -L$_4$ on subpopulations of human T cells was recently investigated (Borrebaeck et al. 1987). Purified T helper/inducer (CD4$^+$) (> 99 %) and suppressor/cytotoxic (CD8$^+$) (> 95 %) cells were stimulated with various concentrations of the two PHA isolectins. CD4$^+$ but not CD8$^+$ cells were induced to produced IL-2 by both isolectins. This ability to induce IL-2 production was very similar between the two isolectins when the stimulation was performed under serum-free conditions using low lectin concentration (Glad and Borrebaeck 1984). This is in contrast to a report by Hernandez and Leavitt (1984) where they stated that PHA-E$_4$ was a poor inducer of IL-2; the isolectin was, however, used at a concentration of 90 μg · ml^{-1} which is at least 20–30 times too high for an optimal induction of cell proliferation (Glad and Borrebaeck 1984).

The only significant difference in biological activity between the two isolectins was the fact that PHA-E$_4$ induced CD4$^+$ cells to proliferate at only half the rate as compared to PHA-L$_4$ (Fig. 1.4). This is despite the fact that the two isolectins induce a similar IL-2 production in CD4$^+$ cells and it might indicate a difference in the ability of the two isolectins to induce expression of IL-2 receptors. The weak induction of IL-2 in CD8$^+$ cells compared to CD4$^+$ cells was, however, most probably due to inherent properties of PHA since CD8$^+$ cells have been reported to produce IL-2 when stimulated with anti-CD3 antibodies (Welte et al. 1983) or Con A and a phorbol ester (PMA) (Luger et al. 1982).

1.4 Induction of Suppressor Cells

The generation of activated non-specific suppressor T cells by mitogens, particularly Con A, has been described by several authors. These suppressor cells inhibit mitogen or antigen-driven lymphocyte proliferation (Palacios and Möller 1981; Lomnitzer et al. 1984), production of IFN (Torres et al. 1982a, b) and IL-2 (Palacios and Möller 1981; Gullberg and Larsson 1982; Lomnitzer et al. 1984; Carlsson et al. 1987a) and inhibit production of antibodies against T cell dependent antigens by B cells (Palacios and Möller 1981; S. R. Smith et al. 1984). The mechanism(s) underlying these suppressor phenomena is not clear but suppression is not the indirect effect of non-specific killing of the responder cells (Nair and Schwarz 1981). Although most investigators have used co-culture systems in their investigations, some results point to the fact that suppressor activity can be mediated by soluble factors (Fleisher et al. 1981; Greene et al. 1981a; Salinas-Carmona 1982).

However, over the past years there has been some controversy regarding the validity of the concept of mitogen-induced, non-specific suppressor cells. Some authors report that the suppression of lymphocyte functions is an active suppression (Gullberg and Larsson 1982; Lomnitzer et al. 1984; S. R. Smith et al. 1984), whereas others present results that support the concept that the suppression is due to absorption of IL-2 from the culture medium by the IL-2 receptor-bearing, so-called suppressor cells (Palacios and Möller 1981; Torres et al. 1982a; Thoman and Weigle 1984; G. Möller 1985; Carlsson et al. 1987a). Addition of exogenous IL-2 to such suppressor cell cultures reverses the suppression of different lymphocyte functions (Palacios and Möller 1981; Torres et al. 1982). Inhibitory effects of suppressor cells on IL-2 production by fresh lymphocytes can also be abrogated by the addition of monoclonal antibodies which block the absorption of IL-2 from the culture medium (Carlsson et al. 1987a). Furthermore, Möller (1985) has shown that Con A activated lymphocytes with suppressor activity in vitro were helper cells in vivo.

Con A induced suppressor cells have been claimed to belong to the CD8 subset (Gullberg and Larsson 1982; Torres et al. 1982a, b). In contrast, we found that not only CD8[+] T cells but also CD4[+] T cells act as suppressor cells provided they express receptors for IL-2 (Carlsson et al. 1987a).

In conclusion, most of the results on lectin induced suppressor cells, especially from experiments utilizing a co-culture system, should be interpreted with great caution since they might be due to experimental artifacts. However, some aspects of the Con A induced suppressor phenomenon are difficult to explain by increased IL-2 absorption. This is especially true for the inhibition mediated by Con A induced lymphokines as reported by Greene et al. (1981a) and Fleisher et al. (1981). Interestingly, these lymphokines which have been classified as endogenous lectins were suggested to bind to the same surface receptors as WGA and *Agaricus bisporus*. Both these lectins are known to inhibit lymphocyte activation. The molecular nature and mode of action of the suppressive endogenous lectins await further investigation.

1.5   References

Abo T, Miller CA, Balch CM, Cooper MD (1983) Interleukin 2 receptor expression by activated HNK–1[+] granular lymphocytes: a requirement for their proliferation. J Immunol 131:1822–1826

Alcover A, Ramarli D, Richardson NE, Chang H-C, Reinherz EL (1987) Functional and molecular aspects of human T lymphocyte activation via T3-Ti and T11 pathways. Immunol Rev 95:5–36

Angelisova P, Haskovec C (1978) Isolation and chemical characterization of a highly purified phytomitogen from *Phaseolus coccineus* seeds. Eur J Biochem 83:163–168

Barker BE, Farnes P (1967). Mitogenic property of *Wistaria floribunda* seeds. Nature (London) 215:659–660

Basham TY, Waxdal MJ (1975) The stimulation of immunoglobulin production in murine spleen cells by the pokeweed mitogens. J Immunol 114:715–716

Beppu M, Terao T, Osawa T (1975) Photoaffinity labeling of concanavalin A. Preparation of a concanavalin A derivative with reduced valence. J Biochem 78:1013–1019

Bessler W, Resch K, Ferber E (1976) Valency-dependent stimulating effects of lima bean lectins on lymphocytes of different species. Biochem Biophys Res Commun 69:578–585

Boldt DH, MacDermott RP, Jorolan EP (1975) Interaction of plant lectins with purified human lymphocyte populations: binding characteristics and kinetics of proliferation. J Immunol 114:1532–1536

Borrebaeck CAK, Etzler ME (1982) Comparison of mitogenic properties of two carbohydrate binding proteins from the *Dolichos biflorus* plant. FEBS Lett 145:8–10

Borrebaeck CAK, Rougé P (1986) Mitogenic properties of structurally related *Lathyrus* lectins. Arch Biochem Biophys 248:30–34

Borrebaeck CAK, Schön A (1987) Antiproliferative response of human leukemic cells: lectin induced inhibition of DNA synthesis and cellular metabolism. Cancer Res 47:4345–4350

Borrebaeck CAK, Carlsson R, Danielsson L, Glad C (1987) Both PHA-L4 and -E4 induce IL-2 production and proliferation in subpopulations of human T cells (in press)

Brown JM, Leon MA, Lightbody JJ (1976) Isolation of a human lymphocyte mitogen from wheat germ with N-acetyl-D-glucosamine specificity. J Immunol 117:1976–1980

Campbell PA, Hartman AL, Abell CA (1982) Stimulation of B cells but not T cells or thymocytes by a sialic acid-specific lectin. Immunology 45:155–162

Cantrell DA, Smith KA (1984) The interleukin-2 T-cell system: a new cell growth model. Science 224:1312–1316

Carlsson R, Hedlund G, Sjögren HO (1987a) Abrogation of staphylococcal enterotoxin A-induced suppressor cell activity by the anti-Tac monoclonal antibody. Scand J Immunol 25:11–19

Carlsson R, Fischer H, Dohlsten M, Sjögren HO, Lando P (1987b) CD3 dependent and independent activation of T cells with staphylococcal enterotoxin A (submitted)

Ceuppens JL, Meurs L, Baroja ML, Van Wauwe JP (1986) Effect of T3 modulation of pokeweed mitogen-induced T cell activation: evidence for an alternative pathway of T cell activation. J Immunol 3346–3350

Chilson OP, Boylston AW, Crumpton MJ (1984) *Phaseolus vulgaris* phytohaemagglutinin (PHA) binds to the human T lymphocyte antigen receptor. EMBO J 3:3239–3245

Closs O, Saltvedt E, Olsnes S (1975) Stimulation of human lymphocytes by galactose-specific *Abrus* and *Ricinus* lectins. J Immunol 115:1045–1048

Curtain CC, Simons MJ (1972) Lymphocyte mitogens of indigenous Australian plant species. Int Arch Allergy 42:225–235

Danielsson L, Möller SA, Borrebaeck CAK (1986) Factor dependent Concanavalin A activation of B lymphocytes. In: Bøg-Hansen TC, Van Driessche E (eds) Lectins: biology, biochemistry, clinical biochemistry, vol 5. De Gruyter, Berlin (W), pp 347–355

Davis L, Lipsky P (1985) I. Phorbol esters enhance responsiveness but cannot replace intact accessory cells in the induction of mitogen-stimulated T-cell proliferation. J Immunol 135:2946–2952

Davis L, Lipsky P (1986) Signals involved in T cell activation. II. Distinct roles of intact accessory cells, phorbol esters, and interleukin 1 in activation and cell cycle progression of resting T-lymphocytes. J Immunol 136:3588–3596

Debray H, Rouge' P (1984) The fine sugar specificity of the *Lathyrus ochrus* seed lectin and isolectins. FEBS Lett 176:120–124

Dillner-Centerlind M-L, Axelsson B, Hammarström S, Hellström U, Perlmann P (1980) Interaction of lectins with human T lymphocytes. Mitogenic properties, inhibitory effects, binding to the cell membrane and to isolated surface glycopeptides. Eur J Immunol 10:434–442

Dohlsten M, Sjögren HO, Carlsson R (1986) Histamine inhibits interferon-gamma production via suppression of interleukin-2 synthesis. Cell Immunol 101:493–501

Douglas SD, Kamin R, Davis WG, Fudenberg HH (1969) Biochemical and morphologic aspects of phytomitogens, jack bean, wax bean, pokeweed, and phytohemagglutinin. In: Rieke WO (ed) Proc 3rd Annu Leucocyte Cult Conf, pp 607–621

Edelman GM, Wang JL (1978) Binding and functional properties of concanavalin A and its derivatives. III. Interactions with indoleacetic acid and other hydrophobic ligands. J Biol Chem 253:3016–3022

Etzler ME, Gupta S, Borrebaeck CAK (1981) Carbohydrate binding properties of the *Dolichos biflorus* lectin and its subunits. J Biol Chem 256:2367–2370

Etzler ME, Quinn JM, Schnell DJ, Spadoro JP (1986) Characterization of lectins from stems and leaves from *Dolichos biflorus*. In: Shannon LM, Chrispeels MJ (eds) Molecular biology of seed storage proteins and lectins, Waverly, Philadelphia, pp 65–72

Falasca A, Franceschi C, Rossi CA, Stirpe F (1979) Purification and partial characterization of a mitogenic lectin from *Vicia sativa*. Biochim Biophys Acta 577:71–81

Falasca A, Franceschi C, Rossi CA, Stirpe F (1980) Mitogenic and haemagglutinating properties of a lectin purified from *Hura creptitans* seeds. Biochim Biophys Acta 632:95–105

Farnes P, Barker BE, Brownhill LE, Fanger H (1964) Mitogenic activity in *Phytolacca americana* (pokeweed). Lancet II: 1100–1101

Felsted RL, Leavitt RD, Bachur NR (1975) Purification of the phytohemagglutinin family of proteins from red kidney beans *(Phaseolus vulgaris)* by affinity chromatography. Biochim Biophys Acta 405:72–81

Fleischer B (1987) A novel pathway of human T cell activation via a 103 kD T cell activation antigen. J Immunol 138:1346–1350

23

Fleisher TA, Greene WC, Blaese RM, Waldman TA (1981) Soluble suppressor supernatants elaborated by concanavalin A-activated human mononuclear cells. II. Characterization of a soluble suppressor of B cell immunoglobulin production. J Immunol 126:1192–1197

Fraser AR, Hemperly JJ, Wang JL, Edelman GM (1976) Monovalent derivatives of concanavalin A. Proc Natl Acad Sci USA 73:790–794

Glad C, Borrebaeck CAK (1984) Affinity of phytohemagglutinin (PHA) isolectins for serum proteins and regulation of the lectin-induced lymphocyte transformation. J Immunol 133:2126–2132

Gordon LK, Hammil B, Parker CW (1980) The activation of blast transformation and DNA synthesis in human PBL by wheat germ agglutinin. J Immunol 125:814–819

Greene WC, Waldmann TA (1980) Inhibition of human lymphocyte proliferation by the nonmitogenic lectin wheat germ agglutinin. J Immunol 124:2979–2987

Greene WC, Parker CM, Parker CW (1976) Opposing effects of mitogenic and nonmitogenic lectins on lymphocyte activation. Evidence that wheat germ agglutinin produces a negative signal. J Biol Chem 251:4017–4025

Greene WC, Fleisher TA, Waldmann TA (1981a) Soluble suppressor supernatants elaborated by concanavalin A-activated human mononuclear cells. I. Characterization of a soluble suppressor T cell proliferation. J Immunol 126:1185–1191

Greene WC, Fleisher TA, Waldmann TA (1981b) Stimulation of immunoglobulin biosynthesis in human B cells by wheat germ agglutinin I. Evidence that WGA can produce both a positive and negative signal for activation of human lymphocytes. J Immunol 127:799–804

Greene WC, Goldman CK, Marshall ST, Fleisher TA, Waldmann TA (1981c) Stimulation of immunoglobulin biosynthesis in human B cells by wheat germ agglutinin. J Immunol 127:799–804

Gullberg M, Larsson EL (1982) Studies on induction and effector functions of concavalin A-induced suppressor cells that limit TCGF production. J Immunol 128:746–750

Gupta BKD, Chatterjee-Ghose R, Sen A (1980) Purification and properties of mitogenic lectins from seeds of *Lathyrus satirus* Linn (chicken vetch). Arch Biochem Biophys 201:137–146

Hawrylowicz CM, Klaus GGB (1984) Activation and proliferation signals in mouse B cells. IV. Concanavalin A stimulates B cells to leave G_0 but not to proliferate. Immunology 53:703–711

Hernandez DE, Leavitt RD (1984) Mitogenic and mitogenically defective phytohaemagglutinin isolectins stimulate T-cell growth factor (interleukin-2) production and response in fresh and cultured human T lymphocytes. Cell Immunol 86:101–108

Horejsi V, Kocourek J (1978). Studies on lectins XXXVI. Properties of some lectins prepared by affinity chromatography on O-glycosyl polyacrylamide gels. Biochim Biophys Acta 538:299–315

Hume DA, Weidemann MJ (1980) Mitogenic lymphocyte transformation. Elsevier/North Holland Biomed Press, Amsterdam New York

Isakov N, Scholz W, Altman A (1986) Signal transduction and intracellular events in T-lymphocyte activation. Immunol Today 7:271–277

Itoh KA, Tilden AB, Kumagai K, Balch CM (1985) Leu-11[+] lymphocytes with natural killer (NK) activity are precursors of recombinant interleukin 2 (rIL-2)-induced activated killer (AK) cells. J Immunol 134:802–807

June CH, Ledbetter JA, Rabinovitch PS, Martin PJ, Beatty PG, Hansen JA (1986) Distinct patterns of transmembrane calcium flux and intracellular calcium mobilization after differentiation antigen cluster 2 (E rosette receptor) or 3 (T3) stimulation of human lymphocytes. J Clin Invest 77:1224–1232

Kammer GM, Kurrasch R, Scillian JJ (1984) Capping of the surface OKT 3 binding molecule prevents the T-cell proliferative response to antigens: evidence that this molecule conveys the activation signal. Cell Immunol 87:284–294

Kilpatrick DC, Graham C, Urbaniak SJ (1986) Inhibition of human lymphocyte transformation by tomato lectin. Scand J Immunol 24:11–19

Kilpatrick DC, McCurrach PM (1987) Wheat germ agglutinin is mitogenic, nonmitogenic and antimitogenic for human lymphocytes. Scand J Immunol 25:343–348

Kishi H, Inui S, Muraguchi A, Hirano T, Yamamura Y, Kishimito T (1985) Induction of IgG secretion in a human B-cell clone with recombinant IL-2. J Immunol 134:3104–3107

Kolberg J (1978) Isolation and partial characterization of a mitogenic lectin from *Lathyrus odoratus* seeds. Acta Pathol Microbiol Scand 86C:99–104

Kumar A, Moreau JL, Baran D, Théze J (1987) Evidence for negative regulation of T cell growth by low affinity interleukin-2 receptors. J Immunol 138:1485–1493

Ledbetter JA, June CH, Martin PJ, Spooner CE, Hansen JA, Meier K (1986) Valency of CD3 binding and internalization of the CD3 cell-surface complex control T cell responses to second signals: distinction between effects on protein kinase C, cytoplasmic free calcium, and proliferation. J Immunol 136:3945–3952

Lipsick JS, Beyer EC, Barondes S, Kaplan NO (1980) Lectins from chicken tissues are mitogenic for Thy-1 negative murine spleen cells. Biochem Biophys Res Commun 97:56–61

Lis H, Sharon N (1986) Lectins in higher plants. In: Liener IE, Sharon N, Goldstein IJ (eds) The lectins: properties, functions and applications in biology and medicine, Academic Press, London New York pp 265–291

Lomnitzer R, Phillips R, Rabson AR (1984) Suppression of interleukin-2 production by human concanavalin A-induced suppressor cells. Cell Immunol 86:362–370

Lotan R, Lis H, Rosenwasser A, Novogrodsky A, Sharon N (1973) Enhancement of the biological activities of soybean agglutinin by cross-linking with glutaraldehyde. Biochem Biophys Res Commun 55:1347–1355

Lotan R, Lis H, Sharon N (1975) Aggregation and fragmentation of soybean agglutinin. Biochem Biophys Res Commun 62:144–150

Luger TA, Smolen JS, Chused TM, Steinberg AD, Oppenheim JJ (1982) Human lymphocytes with either the OKT4 or OKT8 phenotype produce interleukin-2 in culture. J Clin Invest 70:470–473

McClain DA, Edelman GM (1976) Analysis of the stimulation-inhibition paradox exhibited by lymphocytes exposed to concanavalin A. J Exp Med 144:1494–1508

Meuer SC, Hussey RE, Fabbi M, Fox D, Acuto O, Fitzgerald KA, Hodgdon JC, Protentis JP, Schlossman SF, Reinherz EL (1984) An alternative pathway of T-cell activation: A functional role for the 50 kd T11 sheep erythrocyte receptor protein. Cell 36:897–906

Miller K (1983) The stimulation of human B and T lymphocytes by various lectins. Immunobiology 165:132–146

Möller G (1985) Concanavalin-A-activated lymphocytes suppress immune responses in vitro but are helper cells in vivo. Scand J Immunol 21:31–34

Möller SA, Danielsson L, Borrebaeck, CAK (1986) Concanavalin A induced B cell activation mediated by allogeneically derived helper factors. Immunology 57:387–393

Moretta A, Poggi A, Olive D, Bottino C, Fortis C, Pantaleo G, Moretta L (1987) Selection and characterization of T-cell variants lacking molecules involved in T-cell activation (T3 T-cell receptor, T44, and T11): Analysis of the functional relationship among different pathways of activation. Proc Natl Acad Sci USA 84:1654–1658

Mosmann TR, Coffman RL (1987) Two types of mouse helper T-cell clones. Implications for immune regulation. Immunol Today 8:223–227

Munske GR, Pandolfino ER, Magnuson JA (1981) A comparison of the interactions of the mitogenic and non-mitogenic lima bean lectins with human lymphocytes. J Immunol 127:1607–1610

Nair MP, Schwarz SA (1981) Suppression of natural killer activity and antibody-dependent cellular cytotoxicity by cultured human lymphocytes. J Immunol 126:2221–2229

Novogrodsky A, Katchalski E (1973) Transformation of neuraminidase-treated lymphocytes by soybean agglutinin. Proc Natl Acad Sci USA 70:2515–2518

Novogrodsky A, Lotan R, Ravid A, Sharon N (1975) Peanut agglutinin, a new mitogen that binds to galactosyl sites exposed after neuraminidase treatment. J Immunol 115:1243–1248

Nowell PC (1960) Phytohemagglutinin: an initiator of mitosis in cultures of normal human leukocytes. Cancer Res 20:462–466

O'Flynn K, Russul-Saib M, Ando I, Wallace D, Beverley PCL, Boylston AW, Linch DC (1986) Different pathways of human T-cell activation revealed by PHA-P and PHA-M. Immunology 57:55–60

Palacios R, Möller G (1981) T cell growth factor abrogates concanavalin A-induced suppressor cell function. J Exp Med 153:1360–1365

Pandolfino ER, Namen AE, Munske GR, Magnuson JA (1983) A comparison of the cell-binding characteristics of the mitogenic and nonmitogenic lectins from lima beans. J Biol Chem 258:9203–9207

Presant CA, Kornfeldt S (1972) Characterization of the cell surface receptor for the *Agaricus bisporus* hemagglutinin. J Biol Chem 247:6937–6945

Reed JC, Robb RJ, Greene WC, Nowell PC (1985) Effect of wheat germ agglutinin on the interleukin pathway of human T lymphocyte activation. J Immunol 134:314–323

Reinherz EL, Kung PC, Breard JM, Goldstein G, Schlossman SF (1980) T cell requirements for generation of helper factor(s) in man: analysis of the subsets involved. J Immunol 124:1833–1891

Richardson M, Rougé P, Sousa-Cavada B, Yarwood A (1984) The amino acid sequences of alpha 1 and alpha 2 subunits of the isolectins from seeds of *Lathyrus ochrus* (L) DC. FEBS Lett 175:76–81

Rougé P, Borrebaeck CAK, Richardson M, Yarwood A (1987) Structure-function relationship among *Lathyrus lectins*. Glycoconj J (in press)

Ruddon RW, Weisenthal LM, Lundeen DE, Bessler W, Goldstein IJ (1974) Stimulation of mitogenesis in normal and leukemic human lymphocytes by divalent and tetravalent lima bean lectins. Proc Natl Acad Sci USA 71:1848–1851

Saito M, Takaku F, Hayashi M, Tanaka I, Abe Y, Nagai Y, Ishii S (1983) A role of valency of concanavalin A and its chemically modified derivatives in lymphocyte activation. Monovalent monomeric concanavalin A derivative can stimulate lymphocyte blastoid transformation. J Biol Chem 258:7499–7505

Salinas-Carmona MC, Grey I, Russel P, Nussenblatt RB (1982) Mitogen-induced suppressor factor(s) from human lymphocytes: effects on lymphoid and nonlymphoid cells and biophysical properties. Cell Immunol 71:44–53

Schechter B, Lis H, Lotan R, Sharon N (1976) The requirement for tetravalency of soybean agglutinin for induction of mitogenic stimulation of lymphocytes. Eur J Immunol 6:145–149

Schumann G, Schnebli HP, Dukor P (1973) Selective stimulation of mouse lymphocyte populations by lectins. Int Arch Allergy Appl Immunol 45:331–340

Sela B, Lis H, Sharon N (1973) Isolectins from wax bean with differential agglutination of normal and transformed mammalian cells. Biochim Biophys Acta 310:273–277

Smith KA, Cantrell DA (1985) Interleukin-2 regulates its own receptors. Proc Natl Acad Sci USA 82:864–868

Smith SR, Umland S, Terminelli, C, Watnick AR (1984) A study of the mechanism of Con A-induced immunosuppression in vivo. Cell Immunol 87:147–158

Spits H, Borst J, Peter WT, Capel PJA, Terhorst C, De Vries JE (1985) Characteristics of a monoclonal antibody (WT-31) that recognizes a common epitope on the human T cell receptor for antigen. J Immunol 135:1922–1928

Terao T, Osawa T (1973) Purification of hemagglutinins from *Sophora japonica* seeds by affinity chromatography. J Biochem Tokyo 74:199–201

Thoman ML, Weigle WO (1984) Interleukin-2 induction of antigen-nonspecific suppressor cells. Cell Immunol 85:215–224

Torres BA, Yamamoto J, Johnson HM (1982a) Cellular regulation of gamma interferon production: Lyt phenotype of the suppressor cell. Infect Immun 35:770–776

Torres BA, Farrar WL, Johnson HM (1982b) Interleukin-2 regulates immune interferon (IFN-gamma) production by normal and suppressor cell cultures. J Immunol 128:2217–2219

Toyoshima S, Osawa T, Tonomura A (1970) Some properties of purified phytohemagglutinin from *Lens culinaris* seeds. Biochim Biophys Acta 221:514–521

Toyoshima S, Akiyama Y, Nakano K, Tonomura A, Osawa T (1971) A phytomitogen from *Wistaria floribunda* seeds and its interaction with human peripheral lymphocytes. Biochemistry 10:4457–4463

Trowbridge IS (1973) Mitogenic properties of pea lectin and its chemical derivatives. Proc Natl Acad Sci USA 70:3650–3654

Truneh A, Simon P, Schmitt-Verhulst AM (1986) Interleukin-1 and protein kinase C activator are dissimilar in their effects on IL-2 receptor expression and IL-2 secretion by T lymphocytes. Cell Immunol 103:365–374

Tsoukas CD, Landgraf, B, Bentin J, Valentine M, Lotz M, Vaughan JH, Carson DA (1985) Activation of resting T lymphocytes by anti-CD3 (T3) antibodies in the absence of monocytes. J Immunol 135:1719–1723

Tsuda M (1979) Purification and characterization of a lectin from rice bran. J Biochem (Tokyo) 86:1451–1461

Udey MC, Chaplin DD, Wedner J, Parker CW (1980) Early activation events in lectin-stimulated human lymphocytes: evidence that wheat germ agglutinin and mitogenic lectins cause similar early changes in lymphocyte metabolism. J Immunol 125:1544–1550

Ulmer AJ, Scholz W, Flad H-D (1982) Stimulation of colony formation and growth factor production of human T lymphocytes by wheat germ lectin. Immunology 47:551–556

Van Wauwe JP, De Mey JR, Goossens JG (1980) OKT3: a monoclonal anti-human T lymphocyte antibody with potent mitogenic properties. J Immunol 124:2708–2713

Vasquerz A, Gérard JP, Olive D, Auffredou MT, Dugas B, Karray S, Delfraissy JF, Galanaud P (1986) Different human B-cell subsets respond to interleukin 2 and to a high molecular weight B cell growth factor (BCGF). Eur J Immunol 16:1503–1507

Vilcek J, Henriksen-Destefano D, Siegel D, Klion A, Robb, RJ, Le J (1985) Regulation of IFN-gamma induction in human peripheral blood cells by exogenous and endogenously produced interleukin-2. J Immunol 135:1851–1856

Wang JL, Edelman GM (1978) Binding and functional properties of concanavalin A and its derivatives. I. Monovalent, divalent, and tetravalent derivatives stable at physiological pH. J Biol Chem 253:3000–3007

Wang JL, Becker JW, Reeke GN Jr, Edelman GM (1974) Favin, a crystalline lectin from *Vicia faba*. J Mol Biol 88:259–262

Weiss A, Imboden J, Shoback D, Stobo J (1984) Role of T3 surface molecules in human T-cell activation: T-3 dependent activation results in increased cytoplasmic free calcium. Proc Natl Acad Sci USA 81:4169–4173

Welte K, Platzer E, Wang CY, Kan EAR, Moore MAS, Mertelsmann R (1983) OKT8 antibody inhibits OKT3 induced IL-2 production and proliferation in OKT8$^+$ cells. J Immunol 131:2356–2361

Yachie A, Hernandez D, Blaese RM (1987) T3-T cell receptor (Ti) complex-independent activation of T cells by wheat germ agglutinin. J Immunol 138:2843–2847

Yamaguchi N, Yoshimatsu K, Toyoshima S, Osawa T (1981) Isolation and characterization of a mitogenic substance for murine and human B lymphocytes from *Ulex europeus* seeds. J Immunol 126:2290–2295

Yang SY, Chouaib, S, Dupont B (1986) A common pathway for T lymphocyte activation involving both the CD3-Ti complex and GD2 sheep erythrocyte receptor determinants. J Immunol 137:1097–1100

Yarwood A, Richardson M, Sousa-Cavada B, Rougé P (1985) The complete amino acid sequences of the beta$_1$- and beta$_2$-subunits of the isolectins LoLI and LoLII from seeds of Lathyrus ochrus (L.) DC. FEBS Lett 184:104–109

2 Viscaceae Lectins

Hartmut Franz

2.1 Introduction

For the last 100 years toxic plant lectins consisting of A- and B-chains have been of special interest for the following reasons:

— The most characteristic toxic lectin, ricin, was the first lectin detected 100 years ago by Stillmark and Kobert (for review see Franz 1988a, b; Sharon and Lis 1988). The preparation of ricin represents the beginning of lectin research. Also the early development of immunology was essentially influenced by the experiments of Paul Ehrlich with ricin and abrin.
— Toxic lectins are the natural prototype of the so-called immunotoxins. In these conjugates the B-chain (the lectin component) is substituted by antibodies with high specificity of binding to distinct markers of target cells.
— The toxic lectins known so far have very interesting and in some respects even surprising biological activities resulting from overlapping of the different properties of their A- and B-chains.
— Ricin and abrin have been used therapeutically mainly against cancer, with more or less success. Mistletoe lectins are contained in mistletoe preparations (Iscador®, Helixor®, Isorel® a. o.) which are used rather widely in tumor treatment (for review see Leroi 1987; Portalupi 1987). The mistletoe lectins therefore belong to the ones most frequently injected into man. On the other hand, the literature about mistletoe lectins is far less extensive than that about ricin and abrin.

Mistletoes are commonly flowering plants belonging to the three plant families Viscaceae, Loranthaceae and Eremolepidaceae within the order Santalales. Mistletoes are hemiparasites (also called water parasites, partial parasites or aerial parasites (Calder 1983). All the mistletoe lectins described so far have been isolated from plants of the Viscaceae family. Our own investigations suggest that mistletoes throughout the world are a rich source of lectins.

2.2 Preparation of Viscaceae Lectins

The first trials to isolate lectins from *Viscum album* (see color plate) have been reviewed (Franz 1985; 1986, Luther and Becker 1986; Baudino and Sallé 1987). Now the most reliable method for the isolation of Viscaceae lectins is affinity chromatography using carriers with D-galactosyl- (or N-acetyl galactosaminyl-)groups like activated Sepharose, lactosyl-Sepharose or IgG-Sepharose (Ziska et al. 1978; Franz et al. 1981) or by means of cross-linked or heat-aggregated glycoconjugates (Franz et al. 1977).

Mistletoe originalpainting by Wilhelm Ostwald

ML I: The main lectin from *Viscum album* was named by us mistletoe lectin I. Later, other authors used the terms VAA (*Viscum album* agglutinin) (Samtleben et al. 1985) or viscumin (Stirpe et al. 1982). The fresh plant material, always from the same host tree, preferably (*Robinia pseudoacacia* and *Betula pendula*) was air dried at room temperature. Then 100 g of the ground material was mixed with 750 ml of 0.15 M NaCl. After centrifugation the supernatant was applied to a column (3 × 30 cm) of acid-treated Sepharose or lactosyl-Sepharose (Franz 1986). An elution using 0.2 M D-galactose and subsequent dialysis against distilled water yields about 40–50 mg ML I from 100 g dried plant material.

ML II: The effluent passing a column of partially hydrolyzed Sepharose was put onto an immunoglobulin-Sepharose column prepared by coupling human immunoglobulin (mainly IgG) to CNBr-activated Sepharose 4B (20 mg · ml^{-1} gel). Corresponding elution by 0.2 M galactose yielded 2–3 mg from 100 g plant material, depending on the original material.

ML III: After displacing ML II the third lectin ML III was eluted by glycin/HCl pH 2.6. The hemagglutinating fractions were pooled and neutralized with Na_2CO_3. The neutral solution was put on a column (1.5 × 30 cm) of Sepharose-N-(6-aminohexanol)β-D-galactosamine. The column was washed and ML III was eluted with glycine/HCl buffer, pH 2.6. The subsequent operations are the same as described for ML I. VAA2 is presumably identical with ML III; yield: 10–20 mg per 100 g dry plant.

Phoradendron californicum, Dendrophthora serpillifolia, Dendrophthora buxifolia: The original material came from California *(Ph. californicum)* or from Cuba *(D. serpillifolia, D. buxifolia).* Dry plants were treated as described for ML I. The yield was 22 mg lectin from 30 g *Ph. californicum* (Franz et al. 1988a), 23 mg from 80 g *D. serpilifolia* and 15 mg from 50 g *D. buxifolia* (Franz et al. 1986). *Dendrophthora cubensis, D. epiviscum, D. exisa* and *D. tetrostachya* (all grown in Cuba) also contain lectins which agglutinate human erythrocytes and cross-react with polyclonal anti-ML I antibodies. In contrast to the lectins prepared from Cuban Dendrophthorae the lectin from *Phoradendron californicum* (PCL) did not react with anti-ML I antibodies. From the investigated Viscaceae only *Phoradendron piperoides* (from Cuba) was apparently lectin-free.

2.3   Preparation of A- and B-Chains

The A- and B-chains of ML I were prepared under non-denaturing conditions according to Figure 2. 1 (Franz et al. 1982). The lectin was dissolved in PBS, pH 7.2 (about 0.2 mg ML I/ml) and applied to a column of partially hydrolyzed Sepharose 4B or agarose. Non-bound proteins were washed out with PBS. Then 2-mercaptoethanol (5 %) was applied to the column. After the appearance of the 2-mercaptoethanol in the effluent the flow was stopped and the reduction of ML I in the column was allowed to proceed overnight at room temperature. Thereafter the column was washed with PBS. The A-chain passed the column together with 2-mercaptoethanol. The B-chain remained bound and was eluted with 0.1 M lactose in PBS. This procedure is based on the fact that only the B-chain has sugar-binding sites. If the A-chain solution contains small amounts of B-chain, the procedure has to be repeated. The B-chain prepared in this way still contains ML I and has to be purified by DEAE-cellulose chromatography (if necessary, by multiple runs). A- and B-chains prepared by this method differ in their amino acid composition (Franz 1986, Table 2.1).

30

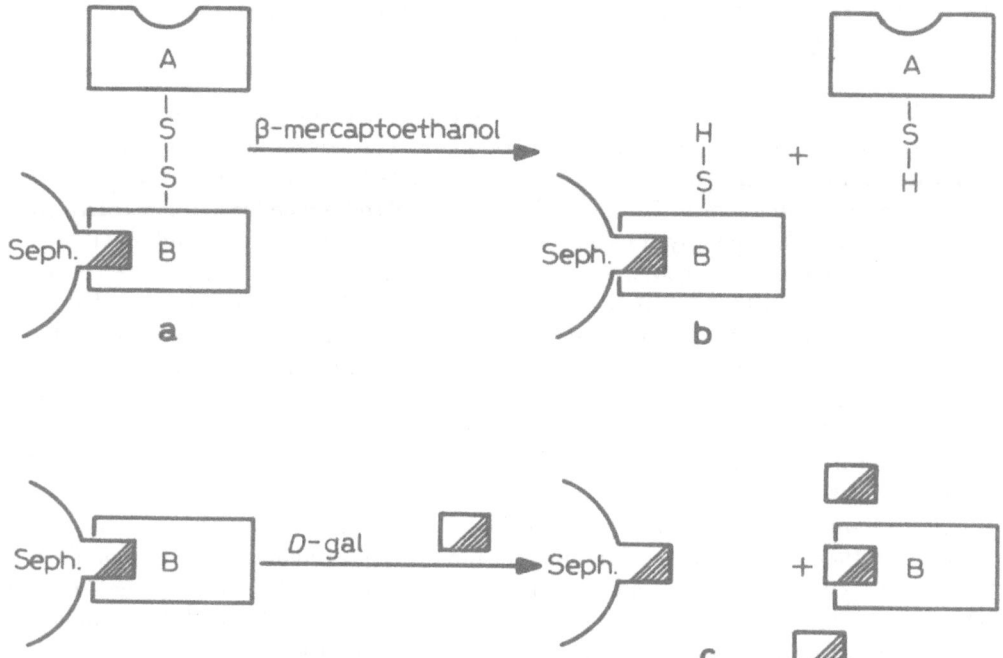

Fig. 2.1 Isolation of A- and B-chains from ML I

a) The lectin ML I is bound to galactosyl groups of the carrier.

b) 2-Mercaptoethanol reduces the S-S bond. The A-chain is released and can be eluated. The B-chain remains carrier-bound.

c) Addition of D-galactose (D-gal) displaces the B-chain.

Seph.: Sepharose 4B

Table 2.1 Amino acid composition of the A and B chains of ML I [mol/mol] (in cooperation with Dr. J. Reinbold, Strasbourg)

	ML I	A-chain	B-chain
Asp	67 (66.63)	24 (24.04)	42 (42.29)
Thr	46 (46.32)	19 (19.22)	26 (26.40)
Ser	44 (44.40)	22 (21.77)	24 (23.70)
Glu	53 (53.02)	29 (28.80)	27 (26.66)
Pro	31 (30.51)	14 (13.97)	15 (15.48)
Gly	49 (49.42)	24 (23.98)	34 (34.24)
Ala	37 (37.13)	17 (17.48)	20 (19.69)
Val	36 (36.26)	14 (13.82)	22 (21.63)
Met	9 (9.14)	4 (3.65)	6 (6.36)
Ile	33 (32.71)	16 (16.05)	17 (16.79)
Leu	49 (49.25)	26 (25.54)	23 (22.45)
Tyr	18 (17.60)	10 (9.98)	7 (7.26)
Phe	19 (18.98)	11 (11.43)	6 (5.59)
His	5 (4.93)	4 (3.53)	2 (1.61)
Lys	8 (8.00)	2 (2.31)	8 (8.17)
Arg	46 (46.00)	24 (23.63)	21 (21.46)
Try	8 (8.10)	1 (0.30)	3 (3.08)
Cys	12 (11.91)	2 (1.77)	11 (10.82)

2.4 Characterization and Quantitative Estimation of *Viscaceae* Lectins

All the Viscaceae lectins described so far consist of two different types of glycoprotein chains named (in analogy to ricin and abrin) A- and B-chains. The B-chain represents the lectin activity, whereas the A-chain is the toxophoric part. A- and B-chains can be demonstrated by polyacrylamide gel electrophoresis (PAGE) in the presence of sodium dodecyl sulfate (SDS) and 2-mercaptoethanol. The majority of A- and B-chains are conjugated by S-S-bonds but there exist also A- and B-chains held together by non-covalent bonds as has been shown by electron microscopy after SDS treatment (Lutsch et al. 1984). These results are in agreement with the findings of Olsnes et al. (1982). ML I tends to form dimers with a molecular weight of about 115,000 MW. The reason for dimerization could be hydrophobic interactions between B- (or A-)chains. Monomeric and dimeric forms of ML I have been electron microscopically demonstrated (Lutsch et al. 1984) as can be seen from Figure 2.2. The ratio of monomers and dimers in aqueous solutions of ML I is concentration-dependent. As shown in Figure 2.3 the higher the ML I concentration, the higher is the percentage of dimers (Lutsch et al. 1984).

The lectin from *Phoradendron californicum* (PCL) has a somewhat higher molecular weight (Franz et al. 1988) than monomer ML I and the other *Viscum album* lectins ML II and ML III. The A- and B-chains of PCL are significantly different from those of ML I as has been shown by SDS-PAGE (Figure 2.4). Also the A- and B-chains of ML I, II and III have different molecular weights and can be distinguished by SDS-PAGE (Franz et al. 1981).

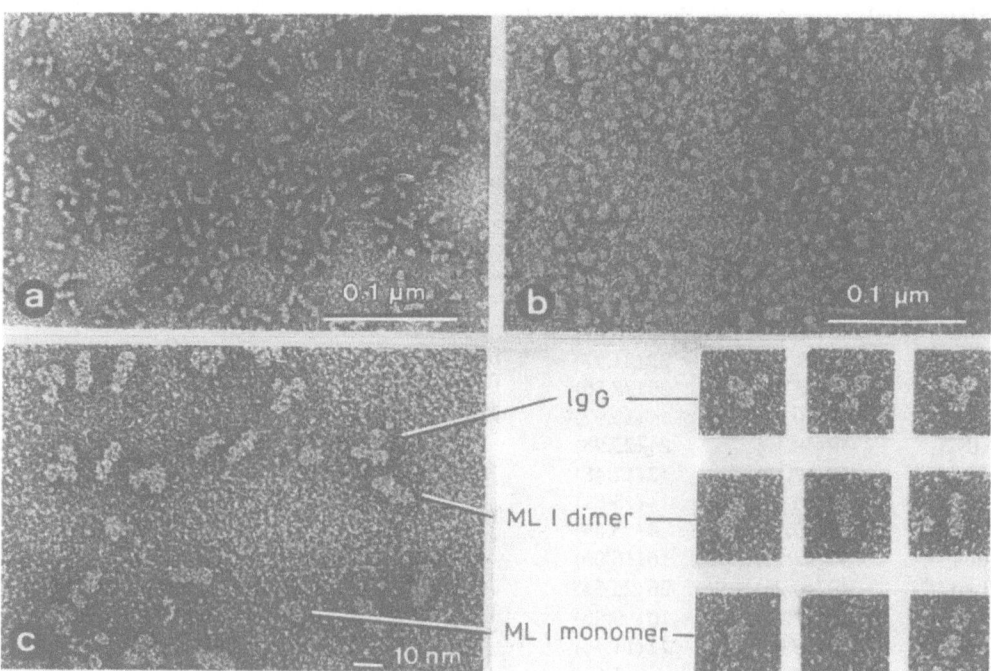

Fig. 2.2 Electron micrographs

a) of negatively stained ML I (200,000 x)

b) of negatively stained ML I after treatment with SDS (200,000 x)

c) of negatively stained mixture of ML I and IgG from rabbit (400,000 x), on the right side selected images

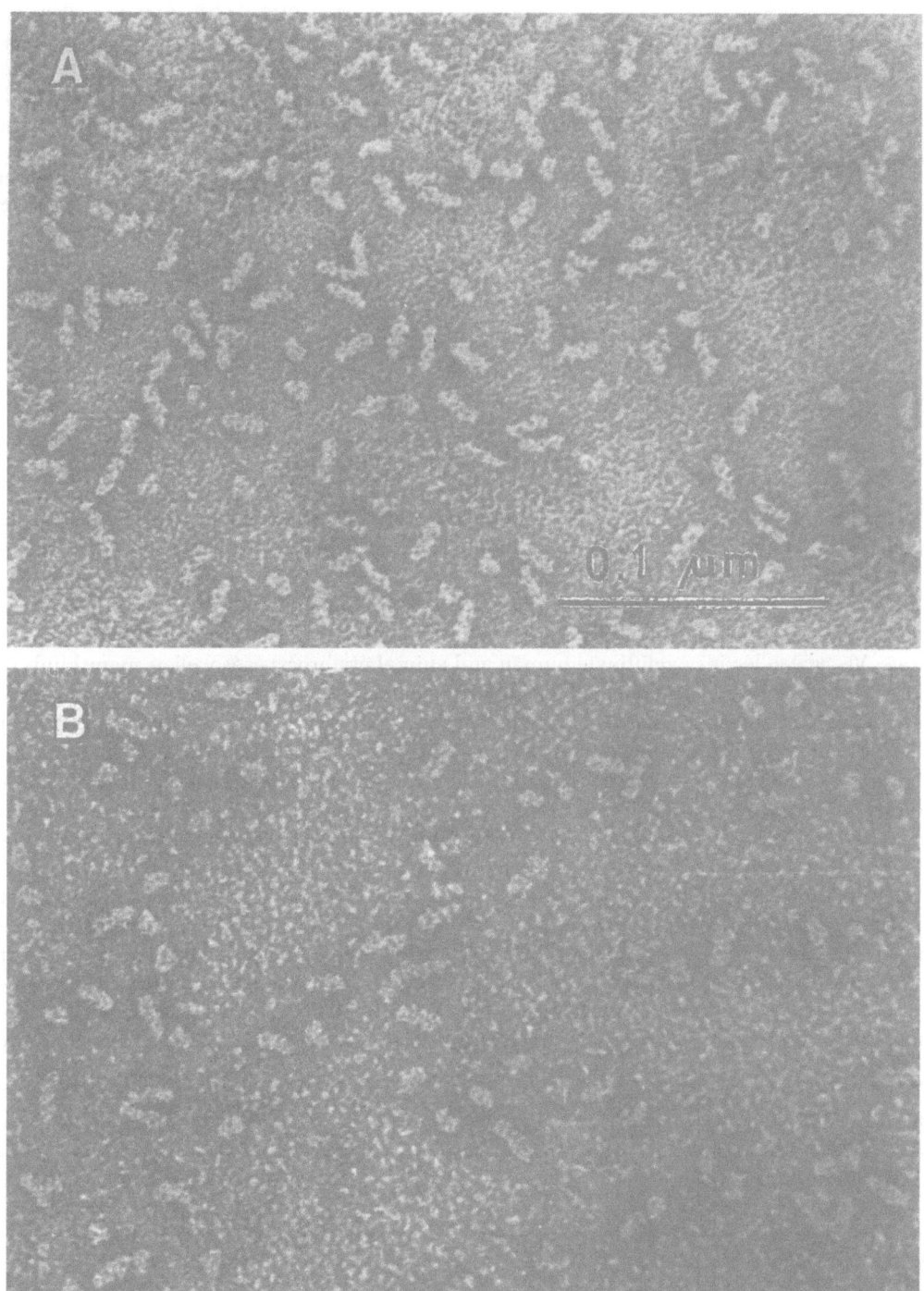

Fig. 2.3 Electron microscope investigation of aqueous solutions of ML I

The sample A) with the higher concentration (40 µg ML I·ml^{-1}) contains more dimers (66 %) than solution B) (4 µg ML I·ml^{-1}, 52 % of dimers).

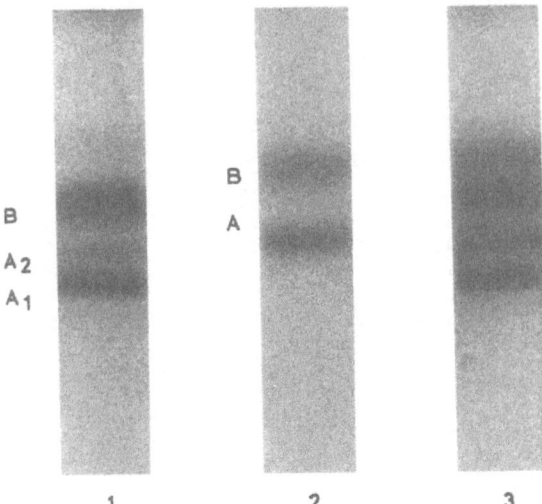

Fig. 2.4 SDS-PAGE of ML I and PCL

1: ML I, 2: PCL, 3: ML I and PCL. Both ML I and PCL consist of A- and B-chains which can be discriminated (3). The A- and B-chains of PCL have a somewhat higher molecular weight. The A-chain of ML I consists of two subtypes (A_1 and A_2). This result is in agreement with investigations together with Walzel (manuscript in preparation).

For the sugar specificity of the B-chains of the different Viscaceae lectins, see Table 2.2. The hemagglutination by ML I and PCL is inhibited by galactosyl derivatives to the same degree (Table 2.3).

The A- and B-chains of the Viscaceae lectins are glycoproteins. The sugar content of ML I was found to be 10.1 (Franz et al. 1981) or 11 % (Luther et al. 1980). PCL contains 14 % carbohydrate. Viscaceae lectins agglutinate human and animal erythrocytes as well as many kinds of other cells (see below) and precipitate glycoconjugates and polysaccharides such as galactomannans from yeast. It is not yet clear whether this agglutination is caused by dimerization of the B-chain or by the existence of two sugar-binding sites per chain (in analogy to ricin). Perhaps a second binding site has a slightly different sugar specificity. Together with other authors we did not find any specificity for human blood groups (Franz 1985). These results are in agreement with those of Samtleben et al. (1985). Only Luther (1974) found in earlier investigations a (weak) blood group B-specificity. For the A-chain, microheterogeneities have been described (Franz 1986). Figure 2.4 demonstrates that the A-chain consists of at least two different species (A_1 and A_2). In contrast to Samtleben et al. (1985), we did not find a corresponding splitting of the carefully purified B-chain in SDS-PAGE. In Table 2.1 the amino acid composition of ML I and its chains is shown (Franz 1986). The differences between A- and B-chains are conspicuous. The solubility of the A-chain in PBS is about $200 \, \mu g \cdot ml^{-1}$. More concentrated solutions form aggregates within a few days. The B-chains have a somewhat higher solubility ($400 \, \mu g \cdot ml^{-1}$).

Table 2.2 Sugar specificity of different Viscaceae lectins

Lectin	MW	Carbohydrate specificity	Blood group specificity	Number of chains	MW of chains
ML I	115,000	D-galactose	none	4	34,000 and 29,000
ML II	60,000	D-galactose/N-acetyl-galactosamine	none	2	32,000 and 27,000
ML III	50,000	N-acetyl-D-galactosamine	none	2	30,000 and 25,000
PCL	68,000	D-galactose	none	2	38,000 and 39,000

34

Inhibitor	Concentration [mM] needed for complete inhibition of 4 hemagglutination units	
	Ph. californicum (PCL)	*Viscum album (ML I)*
D-Gal	25	12
Methyl-α-D-gal	25	12
Methyl-β-D-Gal	25	6
Phenyl-α-D-Gal	25	3
Phenyl-β-D-Gal	12	1.5
6-Deoxy-D-gal	25	12
2-Deoxy-D-gal	> 200	> 200

Table 2.3 Comparison of the inhibitor effect on the hemagglutinating activity of ML I and PCL

The inhibition of protein synthesis at the ribosomal level has been described by Stirpe et al. (1980) and Franz et al. (1982, 1983a). For the quantitative estimation of ML I we developed an ELISA technique (Ziska and Franz 1985). Anti-ML I antibody was diluted to a final concentration of 250 ng protein/ml PBS. Plates were coated with antibody by overnight incubation at 4 °C with 200 µl per well. The antibody was removed, the wells were washed four times with NaCl-Tween 20 (each wash = 3 min) and 200 µl ML I in NaCl-Tween or the probe at an appropriate dilution was added per well. Plates were incubated at 37 °C for 2 h and washed four times with NaCl-Tween. Then 200 µl peroxidase conjugate prepared according to Wilson and Nakane (1978) at a suitable dilution (1 : 1000) in NaCl-Tween 20 (containing 1 % v/v fetal calf serum) was added to each well and incubation continued for 3 h at 37 °C. The plates were washed again. Orthophenylenediamine 4 mg and 15 µl H_2O_2 (30 %) were combined in 10 ml fresh 0.1 M citrate buffer, pH 4.5. The enzyme reaction was allowed to proceed for 20–30 min at room temperature before stopping the reaction with 50 µl of 2.5 N H_2SO_4. The color was read quantitatively on a Titertek Multiscan spectrophotometer at 492 nm. For each determination, a standard curve was generated by plotting the optical density against the concentration of ML I (2.5–20 ng · ml^{-1}; fourfold determinations; Fig. 2.5). The ELISA described above does not allow differentiation between the three lectins ML I, ML II and ML III, because ML II and ML

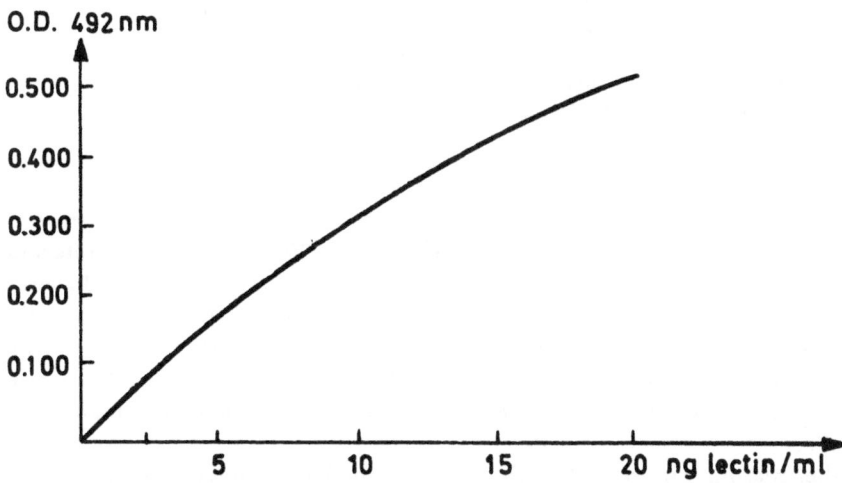

Fig. 2.5 ELISA titration curve of ML I against anti-ML I. Mean absorbance of fourfold determinations of ML I concentrations. O.D.: optical density

Lectin	Concentration [ng · ml⁻¹]	O. D. 492 nm	Corresponds to	[%]
ML I	20	0.430	–	
ML II	20	0.125	2.5 ng ML I/ml	12.5
ML III	20	0.200	5.0 ng ML I/ml	25

Table 2.4 Cross-reaction of ML II and ML III with anti-ML I antibody using the ELISA technique

The cross-reactivity of purified ML II and ML III was expressed relative to ML I.
O. D.: optical density

Month and year of harvest	µg Lectin/g material
November 1982	600
January 1983	1000
March 1983	400
April 1983	450
September 1983	340
October 1983	340
December 1983	480

Table 2.5 Time study of the lectin content of one mistletoe grown on an apple-tree

1 g of dried mistletoe was extracted with 10 ml PBS for 16 h at 4 °C. Proper dilutions of the filtrates were used for lectin determination by the ELISA technique.

Tree	µg Lectin/g material
Crataegus oxyacantha (type of hawthorn)	475
Robinia pseudoaccacia (accacia)	775
Acer saccharinum (maple)	500
Quercus rubra (type of oak)	1250
Populus (poplar)	1800
Tilia (lime tree)	2000

Table 2.6 Lectin content of mistletoe grown on different kinds of trees, harvested on 1 day

1 g of dried mistletoe was extracted with 10 ml PBS for 16 h at 4 °C. Appropriate dilutions of the filtrates were used for lectin determination by the ELISA technique.

III bound to the anti-ML I antibody. ML II and ML III cross-reacted to 12,5 and 25 % respectively in the ML I immunoassay (Table 2.4). Ribéreau-Gayon (1987) found total protection against ML II/ML III cytotoxicity by anti-ML I antiserum (personal communication). However, the total quantity of ML II and ML III in mistletoe extracts amounts only to about 15 % of the total quantity of isolated ML I. We determined the content of lectins in a mistletoe grown on one tree, an apple tree in relation to season (Table 2.5). We observed a seasonal variation, and the content of lectin in mistletoe harvested in the "cold" winter 1982 was greater than in the "warm" winter 1983.

Looking for further reasons with regard to differences in the lectin content, we investigated some mistletoe material grown on different kinds of trees (*Acer, Crataegus, Populus, Quercus, Robinia* and *Tilia*). The plant material was harvested on 1 day (February 1984) in the Great

Component according to Vester (1977)	MW	MW	Component according to Franz et al. (1981)	Table 2.7 Molecular weights of Vester proteins (VP 16)
1	125,000	115,000	ML I (dimer)	
2	70,000	63,000	ML I (monomer)	
3	63,000	60,000	ML II	
4	50,000	50,000	ML III	
5	48,000			
6	33,000	34,000	B-chain of ML I	
7	30,000	30,000	chain of ML III	
8	24,000	25,000	chain of ML III	
9	18,000			
10	14,000			

Garden in Dresden (GDR). We found a fourfold variation. The results are summarized in Table 2.6. Vang (1986) published a modification of the enzyme-labelled lectin assay (ELLA). As the first layer on the microtiter plate he used glycoproteins (asialoorosomucoid or asialofetuin). After binding the lectin reacts with an anti-ML I antibody (rabbit). Finally, an incubation with anti-rabbit-IgG antiserum labelled with horseradish peroxidase is performed. In contrast to the ELISA, this test allows the estimation of carbohydrate binding ML I, whereas the ELISA also recognizes possibly inactivated lectin. This method is, however, less sensitive than the ELISA. The molecular weights of a group of Vester proteins (VP 16) described by Vester in 1977 (Table 2.7) and of lectins from *Viscum album* and their chains respectively, are very similar. Perhaps some fractions of VP 16 are lectins or derived from lectins. Investigating the crude extracts, Vang (personal communication) found a fraction (MW 21,000) reacting with anti-ML I antibodies (blotting technique).

2.5   Modification of ML I

2.5.1   Chemical Modification

The NH_2-, histidyl, tryptophanyl, SH- and COOH-groups of ML I can be chemically modified without loss of agglutinating or precipitating activity. On the other hand, an alteration of the tyrosyl groups results in a decrease of red blood cell agglutination (Ziska et al. 1979). Later, Pfüller et al. (1986) described further modification experiments. Nitration of ML I (more than four $-NO_2$ groups per molecule) results in significant inactivation. Also, the reaction of ML I with the onium salt of tosylchloride and 4-dimethylaminopyridine yields inactivated products. This special inactivation is reversible: at pH 8–9 typical lectin properties (agglutinating activity, toxicity) are reconstituted. Gradual functionalization of amino or carboxyl groups of ML I with activated polyethylene glycol (PEG; MW 2,000 and 6,000 respectively) leads to hydrophilized products with preserved lectin activity. On the other hand, hydrophobization by means of dodecenyl (succinic) anhydride and palmitoyl-N-succinimidate results in ML I derivatives with reduced solubility without inactivation. ML I modified in this way could be easily fixed on hydrophobic matrices including the surface of liposomes. In analogy to ricin, the enzymatic activity of the A-chain is blocked after reaction with phenylglyoxal.

2.5.2   Partial Inactivation by Fused Salts

Recent investigations by Pfüller et al. (1988) demonstrated that fused salts (salts that are liquid (molten) at room temperature) can partially inactivate ML I and other toxic lectins. Most suitable has been ethylammonium nitrate EAN (melting point 12 °C). Also butylammonium-thiocyanate (BAT) has been used. After interaction with EAN for 5–15 min and subsequent elimination of ions by dialysis, ML I loses its ability to bind to D-galactose. It shows neither hemagglutination nor binding to activated Sepharose. On the other hand, it is still inhibiting the protein synthesis in a cell-free system. The toxicity to mice after intraperitoneal (i.p.) application is drastically decreased (Table 2.8).

Similar results have been obtained with ricin. The selective inactivation of biomacromolecules by fused salts may be a new method for modulating the activity of biologically reacting substances. For the potential importance of EAN-treated toxic lectins for the preparation of immunotoxins, see below. The partially inactivated ML I is named d ML I.

Declaration	Animal	Application	LD_{50} [$\mu g \cdot kg^{-1}$]	References
Omelotoxin	mouse	i.p.	80	Lutsik (1975)
ML I	mouse	i.p.	28	Franz et al. (1981)
ML II	mouse	i.p.	1.5	Franz et al. (1981)
ML III	mouse	i.p.	55	Franz et al. (1981)
Viscumin	mouse	i.p.	2.4	Stirpe et al. (1982)
PCL	mouse	i.p.	≈ 4000	Franz and Pfüller (1988)
d ML I	mouse	i.p.	≈ 8000	Franz and Pfüller (1988)

Table 2.8 Toxicity of Viscaceae lectins

2.5.3   Influence of pH on the Conformation and Stability of ML I and Its Chains

The influence of the pH on the conformation and stability of ML I and its isolated chains was investigated by Bushueva et al. (1988), who measured the intrinsic fluorescence. Treatment of the lectin with a denaturant (guanidine hydrochloride) or with quenchers (I^-, Cs^+, acrylamide) revealed a different structure at pH 7.0 and 4.0. At pH 4.0 the tryptophan residues become more accessible to quenchers. The positive charge of the surrounding area increases and the protein becomes more stable to the action of denaturant. The structures of the isolated A- and B-chains differ considerably from that of the whole protein. The stability of both chains to the action of guanidine hydrochloride is lower and for the B-chain the accessibility of tryptophan residues for quenchers increases. Differences between the conformations of the isolated chains at pH 7.0 and 4.0 are marked more strongly. Moreover, at pH 4.5 the B-chain undergoes a structural transition which is possibly related to the auxiliary function of the B-chain in the transmembrane transfer of the lectin A-chain or of an immunotoxin containing d ML I.

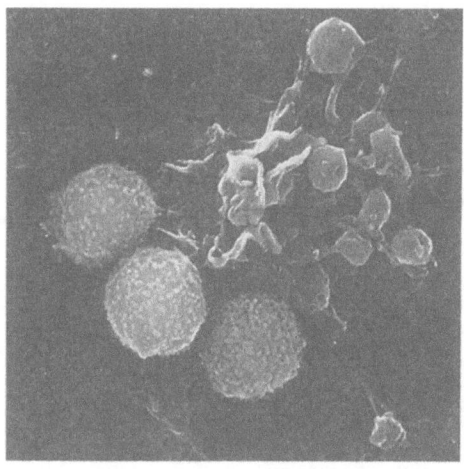

Fig. 2.6 Complex formation of human mononuclear cells after ML I treatment (10^{-8} M). The cell aggregation consists of lymphocytes, monocytes and platelets (4800 x)

2.6 Morphological Studies

Studies together with Neumann et al. (1986) demonstrated that the lectin(s) in *Viscum album* are localized mainly in the vacuoles. The authors demonstrated characteristic lumps which react with gold-labelled anti-ML I antibodies, indicating that they either represent or contain the lectin(s).

Wagner et al. (1978) used ferritin conjugates of ML I for the electron microscopic demonstration of carbohydrates on the cell surface of human erythrocytes and murine tumor cells. Human A_1 erythrocytes showed only a slight local binding of the conjugate. Cells of the mouse ascites tumor strain L 1210 were labelled very tightly on their surface and incorporated ferritin conjugates by pinocytosis. Furthermore, they showed cytotoxic changes in their ultrastructure. In the presence of galactose the labelling on the surface, the incorporation of the conjugate as well as the cytotoxicity were inhibited. Franz et al. (1979) visualized ML I-binding structures on HeLa cells using the system HeLa + ML I + ferritin antibody (the carbohydrate moiety of the antibody is bound by free binding sites of the cell-fixed lectin) + ferritin (bound by the combining sites of the ferritin-specific antibody). This method avoids completely the labelling of the reagents. Using the same procedure, ML I-binding structures on the synaptosomes of the rat cerebral cortex have been pointed out by Ichev et al. (1985). It could be that these synapses differ in the amount of D-galactose-containing glycoconjugates. ML I specifically bound to human red blood cells can also be demonstrated by means of blood group substance A labelled covalently with ferritin. The affinity of isolated Kupffer cells from the rat liver may be important for the elimination of ML I in vivo (Franz et al. 1985). At relatively high concentrations (10 µg · ml^{-1}) ML I destroys these cells in a manner similar to ricin. With human mononuclear cells ML I interacts by the formation of cell complexes consisting of monocytes, lymphocytes and platelets (Metzner et al. 1986). Cell aggregates as shown in Figure 2.6 may influence immunological reactions in vivo.

In 1983 we found, together with Groeger et al., that ML I as well as RCA (*Ricinus communis* agglutinin) most intensively stain microglial cells in cell structures of chick embryo cerebellum. Other cell types remained almost totally unstained. RCA and ML I showed closely related selectivities. The authors were the first to show that D-galactose specific toxic lectins are useful for the cytochemical demonstration of microglia. The same result was found for hepatic cell

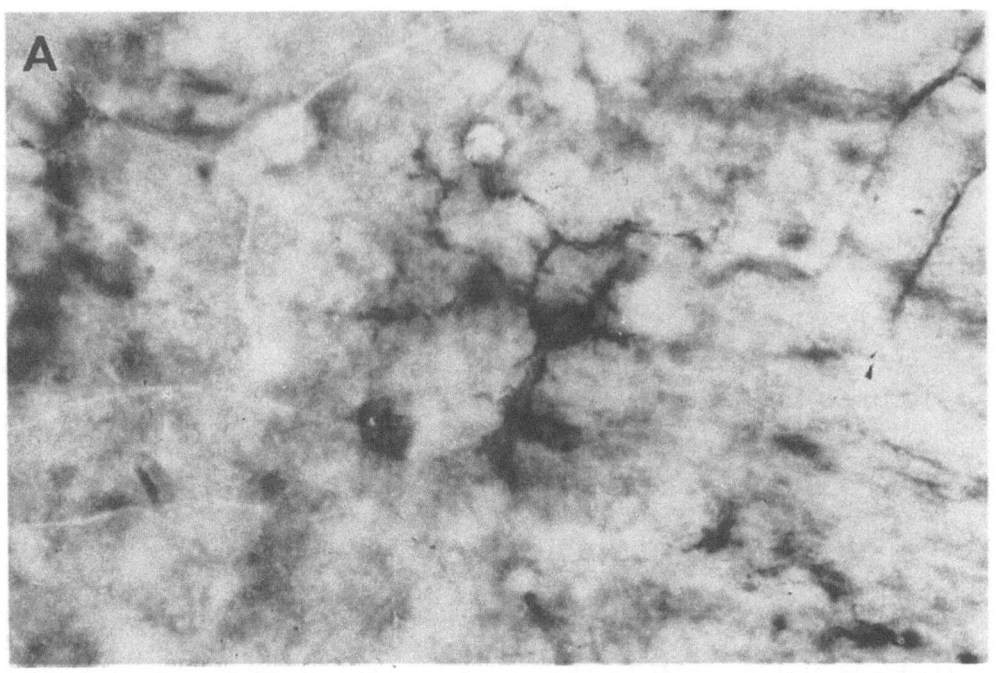

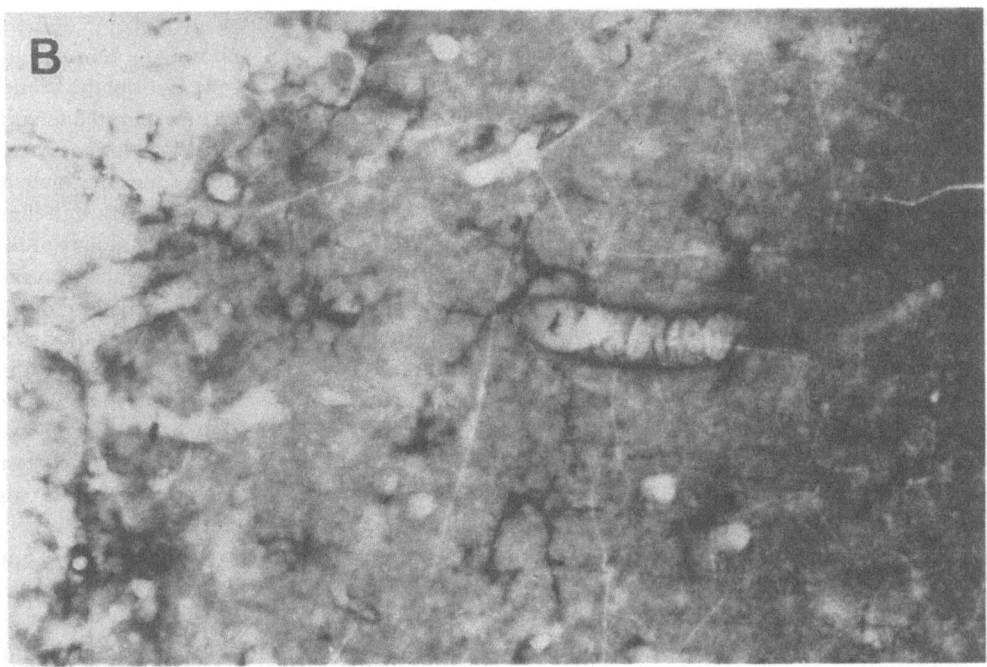

Fig. 2.7 Visualization of microglia in rat cerebellum

A) frozen section (600 x)   B) frozen section (240 x)   C) paraffin section (600 x)

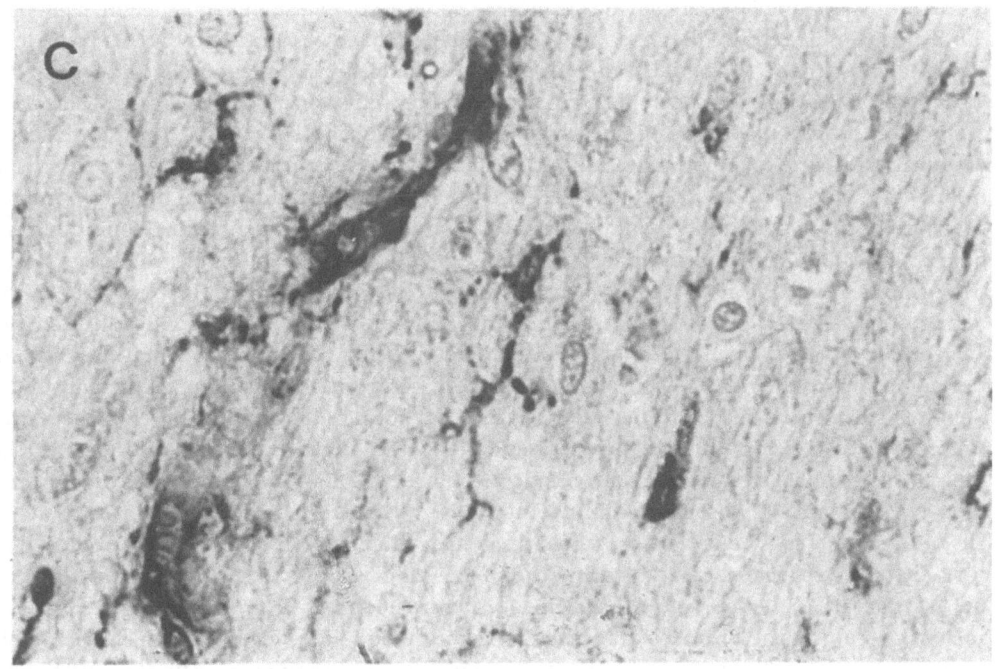

suspensions. Recent studies by the author of this review, together with Suzuki et al. (1988) and with Yamamoto et al. (1988), compared the reactivity of human and rodent microglia with both ML I and RCA I (*Ricinus communis* agglutinin I). Both frozen and deparaffinized sections were incubated for 48 h with either one of the lectins (ML I, RCA I, 3.3 ng · ml^{-1}). After rinsing, the sections were incubated with anti-lectin antibody (from rabbit) for 24 h and then for 1 h with biotinylated anti-rabbit IgG and next with avidin-biotin-HRP complex. After rinsing, the peroxidase reaction was carried out in a common way. Addition of lactose (0.2 M) inhibited the lectin reaction. The authors found a certain quantitative difference of rodent microglia, while RCA I preferentially binds to human microglia but only weakly to rodent microglia.
Figure 2.7 shows microglia in the rat cerebellum after ML I pretreatment.

2.7 Cytotoxic and Anti-Tumor Cell Activity of **Viscaceae** Lectins

2.7.1 In Vitro Experiments

Aqueous extracts of *Viscum album* contain two main groups of cytotoxic substances (for review, see Franz 1985). They differ in their ability to bind to carrier-fixed immunoglobulins (Ig).

- Viscotoxins (MW about 5,000) do not bind. For the different viscotoxins and their properties, see Samuelsson (1973).
- Mistletoe lectins are bound by Ig-Sepharose. It seems that viscotoxins and mistletoe lectins do not share antigenic determinants (Franz et al. 1983b).

In 1975 Lutsik pointed out the obvious similarity between ricin and mistletoe lectin. Together with Stirpe et al. (1980) we demonstrated that the mechanism of cytotoxicity of ML I is strictly related to that of ricin, abrin and modeccin. All these lectins inhibit the protein synthesis at the ribosomal level as has been shown using a lysate of rabbit reticulocytes. The ID_{50} (concentration giving 50% inhibition) was $2.6 \mu g \cdot ml^{-1}$. The effect is enhanced (ID_{50} $0.21 \mu g \cdot ml^{-1}$) if the lectin is reduced with 2-mercaptoethanol. ML I inhibited the protein synthesis also in BL/8L cells in culture after a lag time of 3 h (ID_{50} decreased to $7 ng \cdot ml^{-1}$ after reduction of the S-S-bonds of the lectin). The application of A- and B-chains prepared under non-denaturing conditions enabled Franz et al. (1982) to investigate the influence of the different chains on protein synthesis. Using cell-free rabbit reticulocyte lysate and a rat liver polysomal system, we found that the A-chain (three different preparations) inhibits the (^{3}H)Leu incorporation significantly at the ribosomal level. The effect is strongly dose-dependent. After careful purification the B-chain does not influence the rate of (^{3}H)Leu incorporation. Our results concerning the activity of the A-chain have been confirmed by Olsnes et al. (1982). They pointed out that this chain acts after treatment by SDS on ribosomes. Antiviscumin (anti-ML I) antibodies inhibit this activity. The authors also established a ricinlike enzymatic mechanism. One A-chain molecule of viscumin inactivated at least 50 ribosomes within ten minutes.

The authors concluded that there must be an enzymatic inactivation of the ribosomes caused by the A-chain. The exact enzymatic specificity of A-chains of toxic lectins was unknown till Endo and Tsurugi (1987) demonstrated that ricin, abrin and modeccin (from *Adenia digitata*) attack the 28 rRNA at A 4324 and release adenine from it. Investigations together with Endo and his group (Endo et al. 1988) revealed that ML I also inactivates the ribosomes by cleaving the N-glycosidic bond in 28S rRNA at position A 4324 (for further information, see Endo et al. 1988 and Endo, this Vol.).

Thus, the A-chain of ML I acts as an N-glycosidase, too. Therefore, the so-called toxic lectins can be understood either as lectins containing an additional (toxophoric) A-chain or as N-glycosidases containing a supplementary (haptophoric) lectin chain. In any case the toxic lectins can be defined as naturally occurring conjugates consisting of an N-glycosidase and a D-galactose specific lectin. It is very remarkable that those lectins are involved in plants belonging to completely different families. Perhaps these A-chains are involved in further biological interactions.

Stirpe et al. (1982) found that the sensitivity to viscumin (ML I) of different cell lines differed considerably.

Sargiacomo and Hughes (1982) compared the cytotoxicity of ML I and other toxic lectins. The protein synthesis was found to be inhibited before that of DNA and RNA, indicating that the effect on protein synthesis is responsible for the toxicity of ML I (viscumin) to cells. Cell lines resistant to modeccin and ricin can be fully sensitive to viscumin (ML I). Perhaps these lectins prefer different galactose-containing receptors on the target cell membrane. The presence of Ca^{2+} was necessary for the lectin effect. The sensitivity of cells to viscumin did not vary much between pH 7 and 9. On the other hand, cells were much less sensitive at pH 6.

Ribéreau-Gayon et al. (1986) showed that ML I produced a dose-dependent inhibition of growth of rat hepatoma tissue culture (HTC) and the human T-leukemia cells line Molt 4. Cytostatic effects were obtained at approximately $500 ng \cdot ml^{-1}$ culture medium on HTC cells and only $5 ng \cdot ml^{-1}$ on Molt 4 cells. Further cytotoxic concentrations dependent on cell strains are shown in Table 2.9. Using Molt 4, Doser (1985) described a highly cytotoxic activity of a partially purified mistletoe lectin preparation (Sigma) added to the culture medium. The cytotoxic effect of the lectin(s) was significantly diminished by addition of N-acetyl-galactosamine. On the other hand, D-galactose was without any inhibition effect. Perhaps the cytotoxicity of this

Table 2.9 Cytotoxicity of Viscaceae lectins

Cell strain	Lectin	Cytotoxic Concentration	Remarks	References
Molt 4	ML I	5 ng·ml^{-1}	LD$_{50}$	Ribéreau-Gayon et al. (1986)
Molt 4	ML I (Sigma)	0.048 µg·ml^{-1}	LD$_{50}$	Doser (1985)
Subpopulation M 4 HM 500	ML I (Sigma)	0.467 µg·ml^{-1}	LD$_{50}$	Doser (1985)
Subpopulation M 4 HM 400	ML I (Sigma)	0.391 µg·ml^{-1}	LD$_{50}$	Doser (1985)
HeLa S$_3$	Viscumin (ML I)	100 µg·ml^{-1}	50 % inhibition of protein synthesis	Stirpe et al. (1982)
BHK/C 13	Viscumin (ML I)	2 µg·ml^{-1}	50 % inhibition of protein synthesis	Sargiacomo and Hughes (1962)
Ric 14 (after neuraminidase treatment)	Viscumin (ML I)	35 µg·ml^{-1}	50 % inhibition of protein synthesis	Sargiacomo and Hughes (1962)
Mouse L-fibroblasts	Viscumin (ML I)	0.5 µg·ml^{-1}	50 % inhibition of protein synthesis	Sargiacomo and Hughes (1962)
BL8L	Viscumin (ML I)	0.007 µg·ml^{-1}	50 % inhibition of protein synthesis	Sargiacomo and Hughes (1962) Stirpe et al. (1980)
BL8L	ML I	0.007 µg·ml^{-1}	50 % inhibition of protein synthesis	Sargiacomo and Hughes (1982)
Ric 14	ML I	0.235 µg·ml^{-1}	50 % inhibition of protein synthesis	Ribéreau-Gayon et al. (1986)
Rat hepatoma tissue culture	ML I	500 ng·ml^{-1}	total inhibition of cell growth (2 · 10^3 cells/well)	Franz and Pfüller (1988)
Ehrlich ascites tumor	ML I	2.5 ng·ml^{-1}	total inhibition of cell growth (2 · 10^3 cells/well)	Franz and Pfüller (1988)
	PCL	500 ng·ml^{-1}		
Plasmacytoma P3/X63 Ag 8	ML I	0.018 ng·ml^{-1}	50 % inhibition of ^{3}H-thymidine incorporation	Raabe (pers. commun.)
L 1210	ML I	3.2 ng·ml^{-1}	50 % inhibition of ^{3}H-thymidine incorporation	Khwaja (pers. commun.)
Meth A	ML I	100 ng·ml^{-1}	50 % inhibition of ^{3}H-thymidine incorporation	Jordan (1985)

lectin preparation is mainly due to ML II which, according to H. Wagner and Jordan (1986), is also the essential lectin in the commercial *Viscum album* preparations Iscador®. Doser also isolated Molt 4 subpopulations with minor sensitivity to the lectins(s). The carbohydrate analysis of these cells showed a lower content of N-acetyl-glucosamine, compared with the lectin-sensitive Molt 4 cells. Also, Jordan (1985) found an inhibition of the ^{3}H-thymidine incorporation using Meth A (fibrosarcoma) and L 1210 cells (ED_{50} 0.1 µg ML I/ml and 0,001 µg ML I/ml respectively). The trypan blue test was negative even after treatment of the cells with an ML I solution at a concentration of 10 µg · ml^{-1} for 24 h.

2.7.2   In Vivo Experiments

With regard to the application of Viscaceae lectins for anti-tumor treatment their acute toxicity (see Table 2.8) must be considered. The reason(s) for this extremely high toxicity is not yet clear. The general inhibition of intracellular protein synthesis does not necessarily give a sufficient explanation. In 1975 Lutsik isolated a lectin from *Viscum album* extracts by affinity chromatography using agar gel. He named it omelotoxin [from omela (Russian) mistletoe] and found a molecular weight of 160,000. For several reasons omelotoxin seems to be identical with ML I. A single dose (intraperitoneally) of omelotoxin (10 µg · kg^{-1} body weight of mice) 24 h after the inoculation of Ehrlich ascites tumor cells (EAC) or NK/Ly cells suppressed the growth of ascites. The injection of the lectin(s) 3 or 5 days after the tumor cell inoculation was ineffective. The growth of ascites after inoculation of EAC pretreated with ML I at different concentrations has been investigated by Franz (1986). In the main experiment 10^7 EAC in 0.5 ml of Ringer's solution were incubated with ML I at 37 °C for 60 min. After careful washing (three times) the cells were given intraperitoneally (i.p.) into mice (Table 2.10). Thus 10–300 ng of ML I effect a distinct retardation of tumor growth connected with a characteristic increase in survival of mice. At 300–500 ng the mice appeared to be sick during days 1–5. Surviving animals were without any signs of tumor for at least 20 days. At concentrations higher than the LD_{50} in the incubation medium (660 ng/20 g mouse, i.p. application) the animals died within 1–3 days. The latter effect can be understood in the following way: EAC bind the lectin and act as vehicles. After the

Table 2.10   Effect of EAC pretreatment with ML I on tumor growth in mice

Incubation with	Nanograms per 10^7 EAC in 0.5 ml	Number of mice	Number of tumor-free mice on day					
			3	10	15	20	25	30 after inoculation
ML I	0 (control)	10	10	0				
	10	10	10	10	10	10	8	2
	50	10	10	10	10	6	6	4
	100	10	10	10	10	8	7	6
	200	10	10	7	7	7	7	7
	300	10	7[1]					
	1,000	10	0[2]					
	1,500	10	0[2]					
A-chain	100	10	10	0				
B-chain	100	10	10	0				

[1]: 3 mice died within days 1–3.

[2]: All mice died within days 1–3.

44

cell inoculation the membrane-bound ML I is released by competitive binding to mouse glycoconjugates and causes the death of the animals in the same manner as the original free lectin. Therefore the endocytosis of ML I by EAC during the preincubation is either of very low degree or the internalized ML I is liberated in active form after the death of EAC. The inoculation of a decreased number of pretreated EAC causes an increase of survival. Preincubation with both A- and B-chain solutions (100 ng · 0.5 ml^{-1}) had no influence on the growth of ascites. ML I pretreated EAC were also incubated in mice (300 µg/10^7 cells) by Taubert et al. (1988). Controls received the same number of EAC without any pretreatment. They investigated:

- Body weight of animals in dependence on the time after tumor transplantation.
- Ratio of non-tumor cells to tumor cells evaluated by microscopical differentiation of cell smears from ascite fluid.
- Flow-cytometric analysis of DNA frequency distribution of the intraperitoneal cells.

All control animals died within 19 days after transplantation. The test animals survived at least till day 42. The evaluated cell ratio increased during the first 2 days after inoculation and then decreased up to day 15 in all series including controls. Later, a marked second increase was found in the test animals. The flow-cytometric DNA histograms reflect the course of proliferation and its inhibition in a similar way. The increase of the ratio non-tumor cells/tumor cells can be interpreted as a stimulated defense reaction against the ML I pretreated tumor cells.

In further experiments we treated female ICR mice 24 h after EAC inoculation (10^6 EAC · 0.5 ml^{-1} per mouse) intraperitoneally with ML I. Figure 2.8 shows results representative for a series of experiments. The application of 80 ng ML I (LD$_{50}$ 660 ng) is tolerated by the animals without any complication. Prolongation or survival time is apparent.

Instead of ML I we injected also the isolated A- and B-chains. Because the A- and B-chains are much less toxic, the injected amount could be essentially higher (100 µg of A-chain, 10 µg of B-chain). The survival time is likewise prolonged (see Fig. 2.8). Even after 75 days 20 % of the animals were free of tumor (A-chain). Surprisingly, also the isolated B-chain was effective at concentrations of 5 or 10 µg/animal. Further experiments have to explain whether this activity is caused by the B-chain itself or by impurities, e.g. A-chain of ML I, respectively. Also EAN-treated ML I (i.e. partially denatured ML I with inactivated B-chain but with still active A-chain, named d ML I) inhibits the tumor growth as seen from Figure 2.8.

We found similar effects at d ML I concentrations between 10 and 100 µg (Franz and Pfüller 1988).

Instead of ML I we also injected the much less toxic lectin from *Phoradendron californicum* (PCL). As seen from Figure 2.9 PCL (50 µg/mouse) is active in a comparable manner. The surviving time is drastically increased, too. Similar results were obtained using 100 µg PCL per mouse. Raabe and Storch (1987) investigated the influence of ML I on the growth of the plasmacytoma P3/X63-Ag8 in mice. This tumor produces m IgG$_1$ (K), BALB/c mice received intraperitoneally 10^6–5.10^6 tumor cells in 0.5 ml Eagle medium. The animals were primed with 0.5 ml paraffinum perliquidum. ML I was also given intraperitoneally. For control, mice were given only tumor cells without any further treatment. The growth of the ascites tumor was monitored by controlling the body weight supplemented by electrophoretic estimation of the monoclonal IgG$_1$. Nearly 60 % of the animals were free of tumor 100 days after tumor inoculation when the lectin was applied repeatedly for several days.

Moreover, in a recent experiment (Raabe, personal communication) the majority of mice surviving as a result of ML I treatment were resistant to a second tumor inoculation. In vivo, ML I is already effective after 2 h at a concentration of 0.07 ng·ml^{-1} on the inhibition of ^{3}H-thymidine incorporation in plasmacytoma P3/X63-Ag8 cells (50 %). Recent experiments of

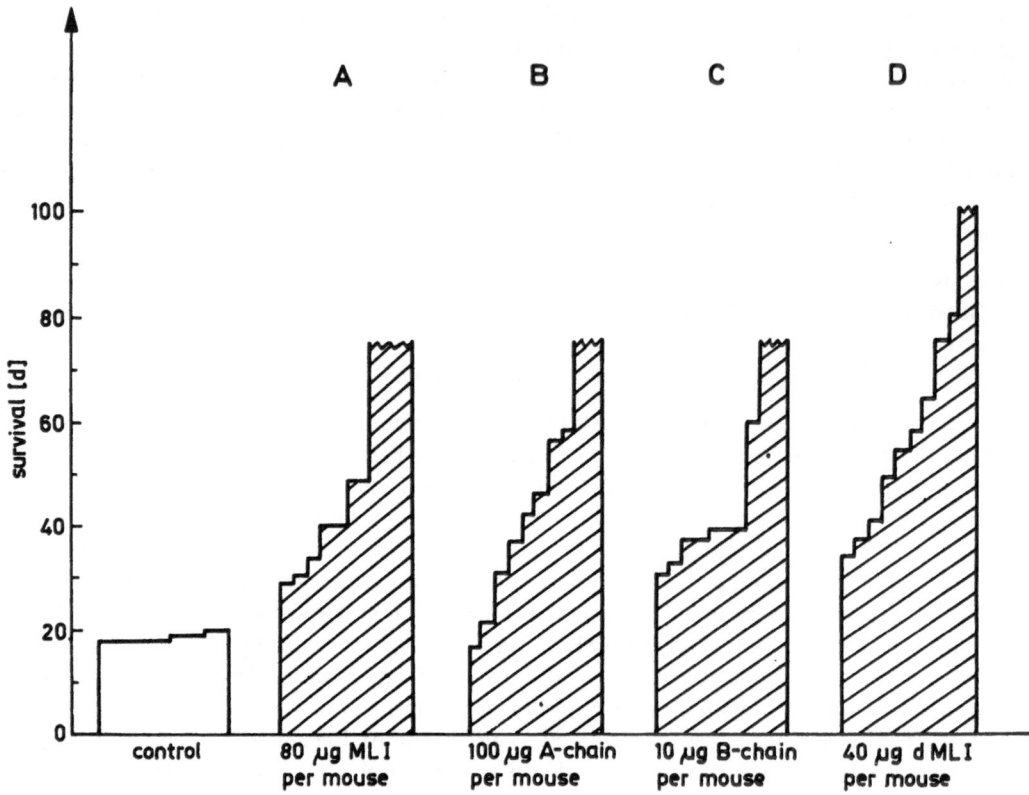

Fig. 2.8 Inhibition of growth of Ehrlich ascites tumor by ML I (A), its A-chain (B), its B-chain (C) and partially inactivated ML I (D)

Groups of 10 mice received 10^5 tumor cells (i.p.) and 6 h later the reagent. Increase of survival is clearly shown. The anti-tumor activity of the B-chain is somewhat surprising. It may be caused by cytotoxic effects, immunostimulating activities and/or macrophage activating influence or by impurification by traces of ML I. Also the activity of non-covalently bound low molecular weight substances can not be excluded.

Raabe (personal communication) revealed a significant inhibition effect even at lower concentration (0.018 ng·ml⁻¹). Stimulated by the experiments of Raabe and Storch, Franz and K. Pfüller also used their plasmacytoma strain. Figure 2.9 demonstrates that these tumor cells are highly sensitive in vivo to subsequent PCL application. After 140 days 12 of 15 animals survived. Application of A- and B-chains of ML I showed corresponding results.

The anti-tumor activity of commercial preparations from *Viscum album* is reviewed here only so far as it is apparently related to the lectin content. Thus, Schröder (1982) investigated the influence of the mistletoe preparation Helixor® on Crown Gall tumors, a plant tumor produced by *Agrobacterium tumefaciens* and found an anti-tumor activity which is inhibited by galactose. Therefore, it is likely that ML I is responsible for this cytotoxicity. Also, other authors (e. g. Havas 1937) observed that certain plant cells are sensitive to mistletoe preparations, but in these cases it was not tested whether lectins were decisive to the cytotoxicity. Very recently, Ribéreau-Gayon (personal communication) found that the serum dilution of a patient who received the ML I-containing preparation Helixor® contains antibodies protecting Molt 4 cells in culture against the toxicity of ML I. These results demonstrate as well the antigenicity of ML I as the

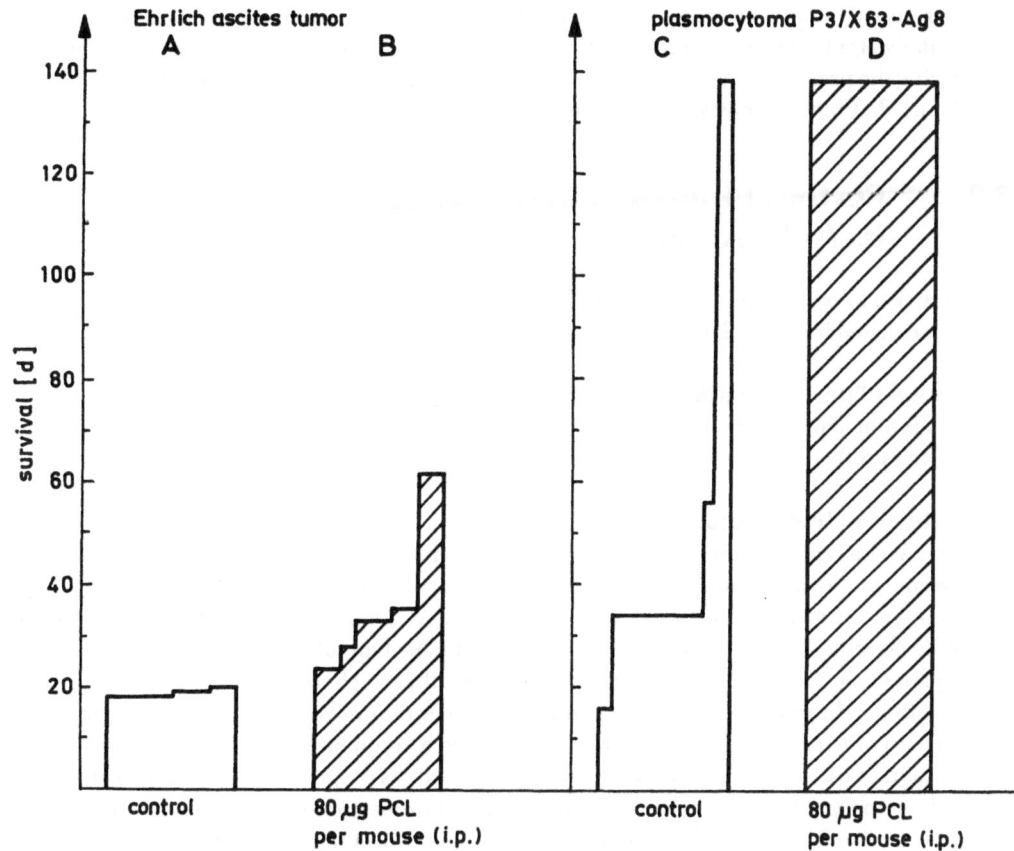

Fig. 2.9 Inhibition of tumor growth after i.p. injection of PCL 6 h following tumor cell implantation is shown

Each group of mice consists of 10 animals. Compared with control groups a characteristic increase of survival time is apparent. The plasmocytoma seems to be much more sensitive to PCL than the Ehrlich ascites tumor.

Remarks:

B: 2 surviving animals were killed after day 61.

C: About 10% of the mice do not develop any tumor after implantation of plasmocytoma.

D: No tumor development in all animals. Mice were killed after day 135.

antitoxic activity of polyclonal anti-ML I antibodies. There is complete agreement with the fundamental ricin experiments of Paul Ehrlich.

Taking into consideration the experimental results reported so far, a final interpretation of the anti-tumor activity of ML I and its chains and of PCL in vivo is very difficult. In any case there are two main mechanisms which should be discussed:

— Immediate cytotoxicity based on the enzymatic activity of the A-chain after binding to the target cell mediated by the haptophoric B-chain.

— Immunostimulating effect mainly by the A-chain but perhaps also by the B-chain. Macrophage stimulation, mitotic effects and lymphokine liberation are under discussion (see also below).

47

Further investigation is necessary to clarify whether fragments of the ML I chains or other low molecular weight substances connected to the *Viscum album* lectins are of biological relevance. Reviewing the present knowledge we do not exclude a unique interference of both mechanisms, which could also be the basis of some therapeutic efforts.

2.8 Further Biological Activities of ML I and Its Isolated Chains

The hitherto known biological activities of these substances, including EAN-treated ML I (d ML I), are listed in Table 2.11.

Inhibition of protein synthesis
The inhibition of protein synthesis caused by the enzymatic activity of the A-chain has been discussed on pp. 60–72

Constituent of immunotoxins
Like the A-chain of ricin (and abrin or diphtheria toxin respectively), the A-chain of ML I can act as a constituent of immunotoxins. Some preliminary results indicate that there is no (or only minimal) immunological cross-reactivity between the A-chains of ricin and ML I. Therefore, it would be possible to use immunotoxin containing the A-chain of ML I even after sensitization to the patient against the A-chain of ricin or against other A-chains.

Meanwhile, preliminary results are accumulating on immunotoxin-like substances containing the A-chain of ML I. Eckert et al. (1985) synthesized a conjugate consisting of histamine and the A-chain of ML I. This conjugate is toxic for histamine receptor-bearing spleen cells of rats (in contrast to the free A-chain and to ovalbumin-histamine conjugates). This conjugate is not an "immuno" toxin because the reaction of the haptophoric component (histamine) with its re-

Table 2.11 Biological activities of ML I and its chains as well as of partially inactivated lectin (d ML I)

Activities	ML I	A-chain	B-chain	d ML I
Inhibition of protein synthesis				
Ribosomal	+	+	−	+
Cellular	+	−	−	−(?)
Constituent of immunotoxins and other				
affinotoxins	−	+	−	+
Mitogenicity	−	+	−	+
Immediate activation of macrophages	+	−	+	not tested
Increase of leukocyte phagocytosis	−	−	+	not tested
Release of macrophage stimulating factor	not tested	+	+	not tested
IL 2 release				
Increase of IgM production	?	+	−	not tested
Inhibition of collagen induced serotonin				
release from platelets	+	−	+	not tested
Inhibition of allergen induced histamine				
release	+	−	+	not tested
Footpad swelling	+	−	−	not tested

+ : positive effect, − : no effect

ceptor has nothing to do with an antigen-antibody reaction. We call such substances "affinotoxins".

The use of EAN-treated toxic lectins may be a new variant for preparing immunotoxins. Utilization of holotoxins consisting of active A-chain and inactivated (i. e. non-D-galactose-binding) B-chain seems to facilitate the membrane penetration by the A-chain. Together with Schütt et al. (1988) we investigated an immunotoxin consisting of a monoclonal antibody against human monocytes (ROMO 1) and d ML I. A suspension of human mononuclear cells has been rendered completely free of monocytes after treatment for 24 h by the corresponding immunotoxin. The elimination of monocytes was confirmed by the non-appearance of lymphocyte stimulation by PHA. Addition of human mononuclear cells treated by the monoclonal antibody alone restores this effect.

Mitogenicity

The A-chain acts as a mitogen (Metzner et al. 1986). It effects a significant stimulation (stimulation index $3 \cdot 5 \pm 1.0$) within a limited concentration range (about 10^{-8} M). Morphologically we also found typical immunoblasts in the A-chain-treated mononuclear cell culture. In contrast, the intact ML I and the B-chain showed a strong inhibitory effect on mitosis, which disappeared with decreasing concentration of ML I and the B-chain. The inhibitory effect of ML I was about 30 times higher than that of the B-chain. It cannot be excluded that the B-chain contains traces of ML I or the A-chain respectively. ML I and its B-chain belong to the group of inhibitory lectins. No cytotoxic effects have been observed using the trypan blue exclusion test. Considering the cytotoxic properties of ML I, we expected such behavior. However, it was very surprising to discover that the A-chain acts as a mitogen, albeit in a narrow concentration range (Fig. 2.10).

Because the A-chain is a glycoprotein, it would be conceivable that it is bound to the lymphocyte surface by membrane lectins. Some years ago, Barzilay et al. (1982) demonstrated mannose-specific membrane lectins on human lymphocytes. If fixation of the A-chain was based exclusively on binding to membrane lectins, one would expect complete inhibition by analogy with

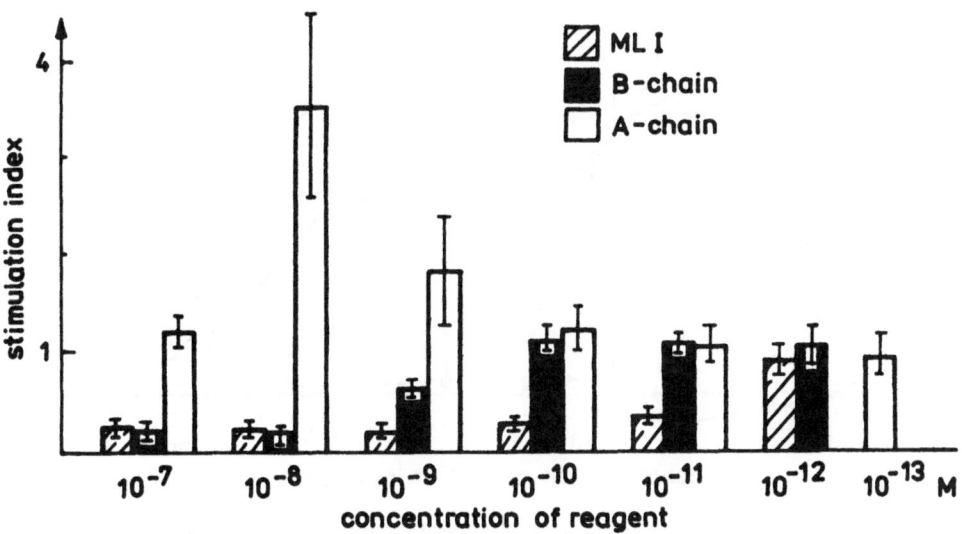

Fig. 2.10 Mitogenic activity of the ML I A-chain within the concentration range of about 10^{-8} M

the Con A/α-methyl-mannoside system. On the other hand, the results of sugar inhibition tests argue against the exclusive role of membrane lectins for binding the A-chain. It is also impossible to exclude the participation of hydrophobic interactions in this phenomenon. The formation of aggregates in aqueous solutions of the A-chain indicates the existence of hydrophobic areas within this molecule. The mechanism of the mitogenic activity of the A-chain is still poorly understood. It would be necessary to test A-chains from other sources (gelonin, A-chain from ricin and abrin) for mitogenic activity. Only partial inhibition was demonstrated with the sugars D-Gal-NH$_2$, D-Glc-NAc and D-Gal-NAc (Metzner et al. 1986) not only for Con A but also for the A-chain. Cell proliferation is a multistep process, in which at least two signals are necessary. Perhaps the sugars tested influence the second (or third) signal.

Probably these findings are in accordance with those of Coeugniet (1985), who described a mitogenic activity of Plenosol® on T-lymphocytes. The mitogenic effect of the ML I A-chain has been confirmed by Kopp (personal communication). It is worthy of note that also d ML I has a mitogenic activity (Metzner; Kopp, personal communications). Hajto and Lanzrein (1986) and Hajto and Hostanska (1986) found an increase of NK cells and ADCC after intravenous application of Iscador®. Hamprecht et al. (1987) investigated Iscador® for its potency to influence NK cytotoxicity in vitro and found a drastic enhancement. One responding effector cell was identified as a member of the large granular lymphocyte family.

Immediate activation of macrophages

ML I has no cytotoxic effect on human granulocytes and paraffin oil stimulated macrophages from guinea pig between 10^{-14} and 10^{-8} M (trypan blue and ethidium bromide exclusion) (Metzner et al. 1985). Similar results have been reported by Luther et al. (1973). Over the same

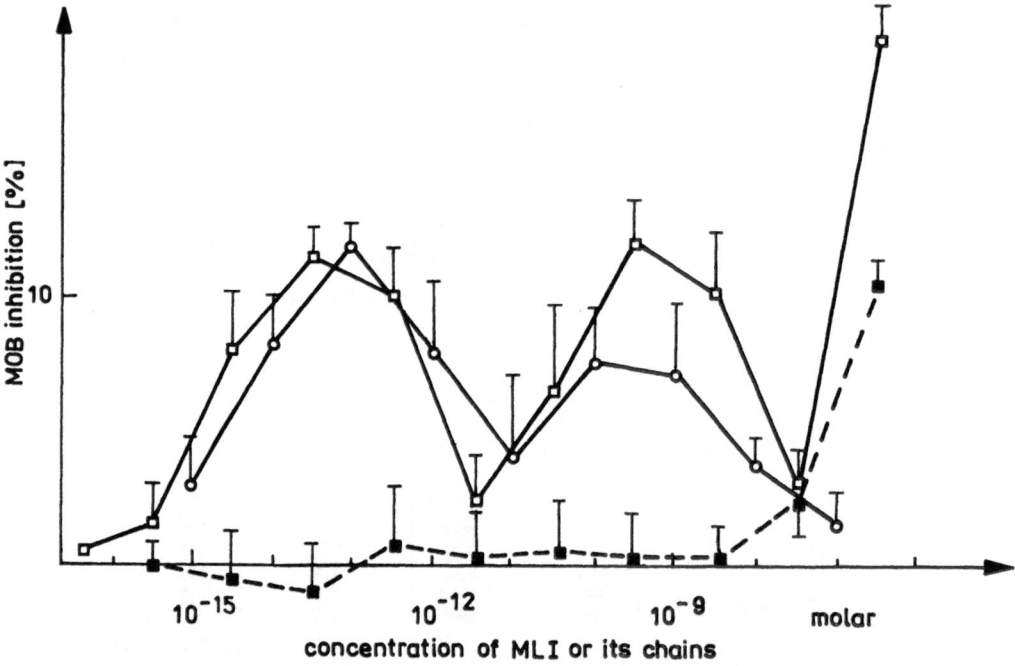

Fig. 2.11 Influence of ML I (o) and its isolated A-(■) and B-chains (□) on the electrophoretic mobility (MOB) of guinea pig macrophages. The main values with standard deviation are shown for ML I, B-chain and A-chain

50

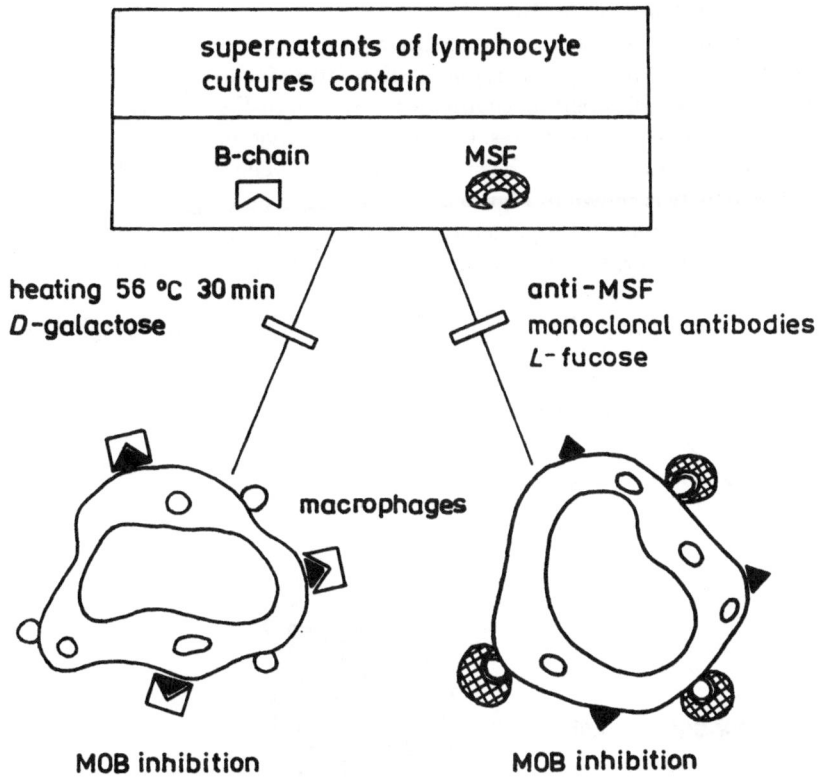

Fig. 2.12 The two macrophage-stimulating activities (shown in Fig. 2.13) are differentiated. The activation by the B-chain is inhibited by heating or by D-galactose. On the other hand, the lymphokine is inhibited by anti-lymphokine monoclonal antibodies or by L-fucose respectively

MSF: macrophage stimulating factor
MOB: electrophoretic mobility

concentration range, ML I and B-chain diminish the negative surface charge of macrophages (electrophoretic mobility test) and agglutinate these cells at concentrations of 2×10^{-8} M (ML I) and 3×10^{-7} M (B-chain) respectively (Metzner et al. 1985).

Diminishing the negative surface charge shows two peaks, indicating the existence of two types of receptors on the macrophage surface with different affinities for the specific binding sites of the lectin. In comparison, the A-chain reduces the surface charge at concentrations over 3×10^{-7} M (Fig. 2.11). In contrast to ML I and B-chain, D-galactose has no effect on this.

Increase of phagocytosing capacity

Only the B-chain increased the phagocytosing capacity of normal human leukocytes against yeast particles *(Saccharomyces cerevisiae)*. ML I and the A-chain caused no alteration of the phagocytic activity. It must be stressed that the concentration of ML I (5×10^{-8} M) was in the agglutinating range. The phagocytosis may therefore be influenced by agglutination of the leucocytes (Metzner et al. 1986). Jordan (1985) described for the same system a small increase (18 %) of phagocytosis after interaction of ML I at a lower concentration (10 ng·ml^{-1}). Still lower concentrations were without any significant effect.

Release of macrophage stimulating factor

Besides the immediate binding (see above), the B-chain can also indirectly activate macrophages by release of lymphokines (e.g. macrophage stimulating factor) from lymphocytes (Metzner et al. 1987). These two different mechanisms are shown in Figure 2.12. Using the supernatants of the B-chain-treated lymphocytes, we found a clear macrophage activation (inhibition of macrophage cell electrophoresis). The differentiation between the direct B-chain effect and the lymphokine activity is shown in Figure 2.13. The sugar-binding activity of the B-chain is heat-sensitive and is inhibited by D-galactose, whereas the lymphokine does not act in the presence of L-fucose. Moreover, the lymphokine is activated by specific monoclonal antibodies.

Thus, the B-chain both inhibits the proliferation of lymphocytes and liberates lymphokines from the same cells.

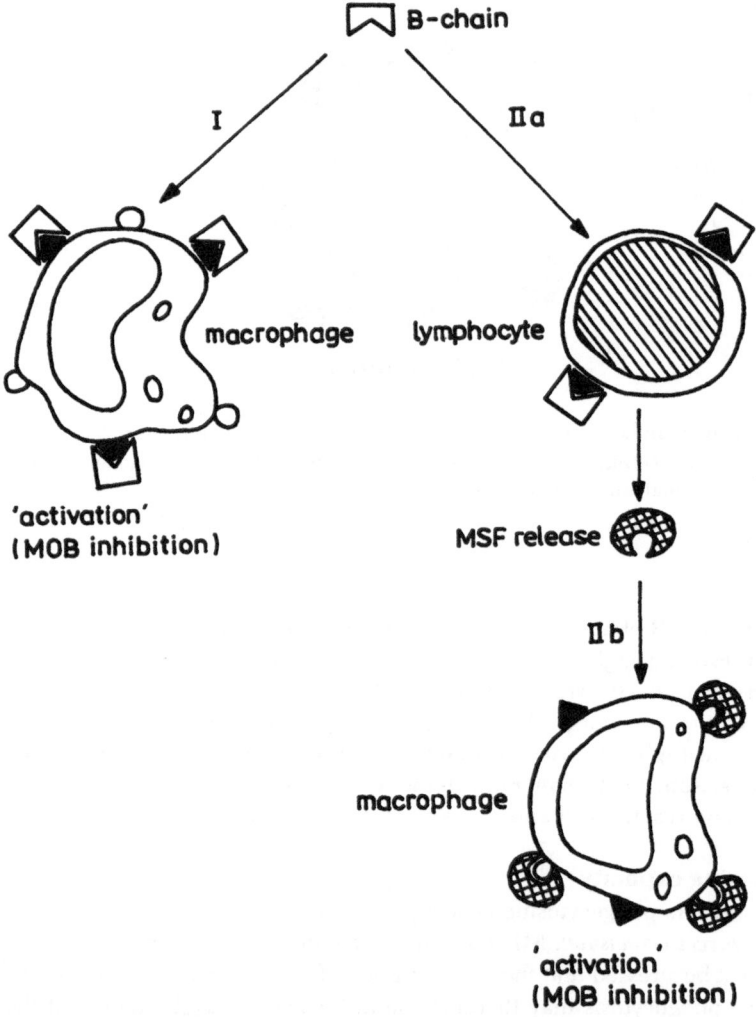

Fig. 2.13 The B-chain activates macrophages immediately (I) and releases (IIa) lymphokines. MSF (macrophage stimulating factor) activates macrophages (IIb). The macrophage activation is represented by decrease of the electrophoretic mobility (MOB)

Table 2.12 IL 2 induction by ML I and its A-chain

Molar concentration of the reagent (ML I or A-chain)	1 Lymphocytes + ML I	2 ML L	3 Lymphocytes + A-chain	4 A-chain
	CTL [cpm]	CTL [cpm]	CTL [cpm]	CTL [cpm]
0	196 ± 21	not tested	196 ± 21	not tested
10^{-5}	148 ± 2	135 ± 21	not tested	not tested
10^{-6}	145 ± 18	136 ± 7	**2128 ± 653**	195 ± 35
10^{-7}	142 ± 11	176 ± 47	**7354 ± 202**	237 ± 23
10^{-8}	158 ± 31	203 ± 41	**1013 ± 132**	254 ± 5
10^{-9}	180 ± 30	181 ± 32	194 ± 4	268 ± 14
10^{-10}	170 ± 0	187 ± 27	188 ± 15	200 ± 4
10^{-11}	179 ± 19	236 ± 31	209 ± 47	188 ± 6
10^{-12}	238 ± 9	358 ± 51	158 ± 6	254 ± 70
10^{-13}	not tested	not tested	199 ± 24	296 ± 3

The ^{3}H-thymidine incorporation in cytotoxic T cells (CTL) after interaction with supernatants (columns 1 and 3) and with the reagents alone (columns 2 and 4) was estimated.

Interleukin 2 (IL 2)-release

IL 2-release from human lymphocytes by the A-chain (10^{-8}–10^{-6} M) was estimated by measuring the ^{3}H-thymidine incorporation in cytotoxic T-cells (CTL). The complete lectin is not effective (Table 2.12). It can be assumed that the IL 2-release is connected with the mitogenic activity of the A-chain. The observed IL 2 induction seems to be different from that of mitogenic lectins. These results point to an immunostimulating activity of the A-chain (Franz et al. 1988b).

Increase of IgM production

The A-chain produces a weak but significant increase of the IgM production in the lymphocyte culture (Kießig, personal communication). ML I is ineffective because of its cytotoxicity.

Effects of ML I on human platelets

Because platelets are involved in numerous immunological and inflammatory processes we (Metzner et al. 1986) investigated their interaction with ML I. Platelets suspended in blood plasma are not agglutinated by ML I at concentrations between 1 to 100 µg·ml^{-1}. On the other hand, washed platelets form agglutinates with the lectin at the same concentrations (Fig. 2.14). This effect can be explained by the competition of plasma proteins which have a D-galactose-containing sugar moiety. Erythrocytes show the same behavior (Metzner et al. 1988). The interaction of ML I with plasma proteins is surely very important for the ML I reactivity in vivo. Only cells with high affinity (i. e. the affinity of these cells to ML I is higher than that of plasma proteins) react with ML I. By this mechanism low affinity cells would be protected against the cytotoxic activity of injected ML I. The agglutination of platelets by ML I is not followed by a serotonin release (in contrast to collagen). The agglutination by the B-chain alone is somewhat weaker. The A-chain is inactive. The platelet agglutination by ML I is strictly suppressed by D-galactose. Serotonin release by collagen is inhibited by a pretreatment of platelets with ML I.

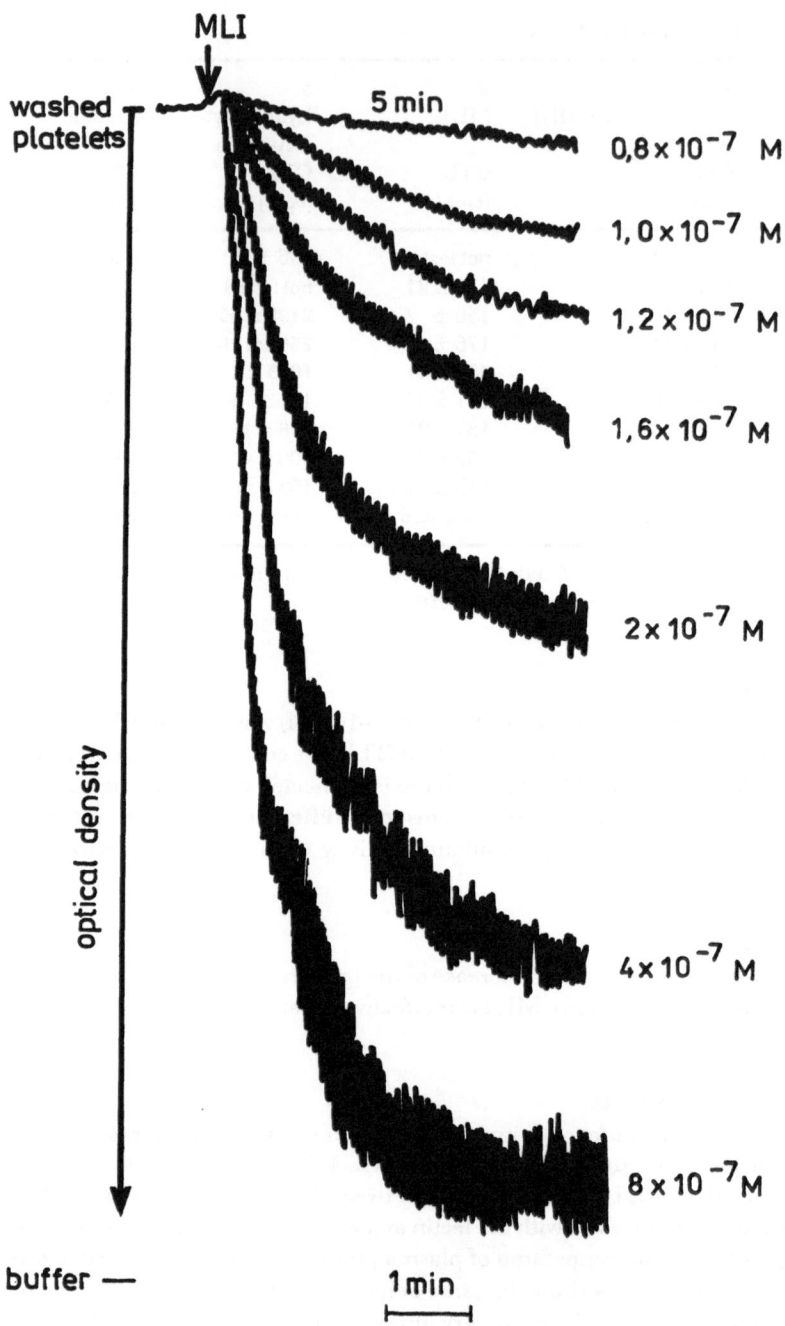

Fig. 2.14 Densitometric measurement of the agglutinating activity of ML I at different concentrations. The agglutination is strongly concentration-dependent and is completed after 5 min

Inhibition of allergen-induced histamine release

The latter finding may be in accordance with the results of Sehrt et al. (1985). They described the inhibition of allergen-induced histamine release from basophil granulocytes by ML I. Likewise the complete lectin and its B-chain, but not the A-chain, were active. The authors assume that the galactose-specific lectins bind to the carbohydrate moiety of the Fc portion of IgE, resulting in a block of interaction of the allergen with antigen-binding sites of IgE. On the other hand, it cannot be excluded that the binding of the B-chain to D-galactose groups on the cell membrane changes the permeability for substances like histamine and serotonin.

Foodpad swelling

ML I produces a characteristic swelling after injection into the footpad of mice (Walzel et al. 1982). A dose-dependent effect was measured in the range of 0.0025–2.5 µg ML I per mouse. This effect can be used as an assay for ML I. Both A- and B-chains are without any activity.

2.9   Final Remarks

In comparison with ricin and abrin there exists a certain demand concerning the investigation of Viscaceae lectins.

It has been the aim of this review to promote the interest in these substances. No one can exclude that Viscaceae lectins or their derivatives are potential drugs.

The author of this review hopes for progress in this field over the next years. Moreover, he is anticipating it. We intend to report on further development in the subsequent volumes of "Advances in Lectin Research".

2.10   References

Barzilay M, Monsigny M, Sharon N (1982) Interaction of soybean agglutinin with human peripheral blood lymphocyte subpopulations: evidence for the existence of a lectin-like substance on the lymphocyte surface. In: Bøg-Hansen TC (ed) Proc 4th Lectin Meet, Copenhagen 1981. Lectins: biology, biochemistry, clinical biochemistry, vol. 2. De Gruyter, Berlin (W), pp 67–81

Baudino S, Sallé G (1986–1987) Les substances actives du gui. Propriétés pharmacologiques et applications thérapeutiques. Ann Sci Nat Bot Paris Ser 13 8:45–72

Bushueva TL, Tonevitsky AG, Maisuryan NA, Kindt A, Franz H (1988) The effect of pH on the conformation and stability of the structure of toxic protein mistletoe lectin I. In: Kocourek J (ed) Lectins: biology, biochemistry, clinical biochemistry, vol 7. Sigma Chemical Company, St. Louis, Missouri, USA (in press)

Calder DM (1983) Mistletoes in focus. An introduction. In: Calder M, Bernhard P (eds) The biology of mistletoes. Academic Press, London New York, pp 1–18

Coeugniet E (1985) Viscum album-Extrakte als immunstimulierendes Mittel. Erfahrungsheilkunde 34:104–109

Doser C (1985) Einfluß eines Gesamtextraktes und eines Lektins aus Viscum album L. auf menschliche transformierte Zellen des hämatopoietischen Systems in vitro. Thesis, Hohenheim

Eckert R, Pfüller U, Kindt A, Reichelt E, Franz H (1985) Histaminrezeptor-tragende Lymphozyten. 3. Suppression von Immunreaktionen nach Abtöten Histaminrezeptor-tragender Lymphozyten durch ein Konjugat aus Histamin und der A-Kette des Mistellektins I. Biomed Biochim Acta 44:1239–1245

Endo Y (1989) Mechanism of action of ricin and related toxic lectins on the inactivation of eukaryotic ribosomes. In: Franz H (ed) Advances in Lectin Research, vol 2. Volk und Gesundheit, Berlin, pp

Endo Y, Tsurugi K (1987) RNA N-glycosidase activity of ricin A-chain. Mechanism of action of the toxic lectin ricin on eukaryotic ribosomes. J Biol Chem 262:8128–8130

Endo Y, Tsurugi K, Franz H (1988) The site of action of the A-chain of mistletoe lectin I on eukaryotic ribosomes: the RNA N-glycosidase activity of the protein. FEBS Lett 231:378-380

Franz H (1985) Inhaltsstoffe der Mistel (*Viscum album* L.) als potentielle Arzneimittel. Pharmazie 40:97–104

Franz H (1986) Mistletoe lectins and their A and B chains. Oncology Suppl 1 43:23–34

Franz H (1987) Wirkstoffe aus der Mistel (*Viscum album* L.) von potentieller pharmakologischer Bedeutung. In: 1st Symp Exp Chemother, 23.–27. November, Weimar

Franz H (1988a) The ricin story. In: Franz H (ed) Advances in lectin research, vol 1. Volk und Gesundheit, Berlin, pp 10–25

Franz H (1988b) Hundred years of ricin. In: Freed D, Bøg-Hansen TC (eds) Lectins: biology, biochemistry, clinical biochemistry, vol 6. Sigma Chemical Company, St. Louis, Missouri, USA, pp 7–13

Franz H, Pfüller K (1988) Manuscript in preparation

Franz H, Haustein B, Luther P, Kuropka U, Kindt A (1977) Isolierung und Charakterisierung von Inhaltsstoffen der Mistel (*Viscum album* L.). 1. Affinitätschromatographie von Mistelrohextrakt an fixierten Plasmaproteinen. Acta Biol Med Ger 36:113–117

Franz H, Bergmann P, Ziska P (1979) Combination of immunological and lectin reactions in affinity histochemistry: proposition of the term affinitin. Histochemistry 59:335–342

Franz H, Ziska P, Kindt A (1981) Isolation and properties of three lectins from mistletoe (*Viscum album* L.). J Biochem 195:481–484

Franz H, Kindt A, Ziska P, Bielka H, Benndorf R, Venker L (1982) The toxic A-chain of mistletoe lectin I. Acta Biol Med Ger 41:K9–K16

Franz H, Kindt A, Ziska P, Bielka H, Benndorf R, Venker L (1983a) Further investigations on mistletoe lectin I (ML I): effect of A-chain on cell-free protein synthesis. In: Bøg-Hansen TC, Spengler GA (eds) Proc 5th Lectin Meet, Bern 1982. Lectins: biology, biochemistry, clinical biochemistry, vol 3. De Gruyter, Berlin (W), pp 645–651

Franz H, Kindt A, Eifler R, Ziska P, Benndorf R, Junghahn I (1983b) Differences in toxicity and antigenicity between mistletoe lectin I and viscotoxin A 3. Biomed Biochim Acta 42:K21–K25

Franz H, Steffan AM, Kirn A (1985) Interaction of mistletoe lectin I with Kupffer cells and endothelial cells of mouse liver. Acta Histochem 78:123–129

Franz H, Leiva A, Ziska P (1986) Lectins from Cuban Viscaceae. In: Int Meet *Viscum album*, 9.–12. April, Heidelberg

Franz H, Müller P, Kindt A, Ziska P (1988a) Viscaceae lectins: A new lectin from *Phoradendron californicum*. In: Freed D, Bøg-Hansen TC (eds) Lectins: biology, biochemistry, clinical biochemistry, vol 6. Sigma Chemical Company, St. Louis, Missouri, USA, pp 239–297

Franz H, Friemel H, Buchwald S, Pantikow A, Kopp J, Körner IJ (1988b) The A-chain of lectin I from European mistletoe *(Viscum album)* releases IL-2. IUB-Meeting Prague 1988, (in press)

Groeger BK, Williams LG, Pigott J, Ziska P, O'Dell DS, Williams DJ, Franz H, Debbage PL (1983) Affinities of ricin 120 and mistletoe lectin I for membrane components of liver and nervous tissue. In: Bøg-Hansen TC Spengler GA (eds) Proc 5th Lectin Meet, Bern 1982, Lectins: biology, biochemistry, clinical biochemistry, vol 3. De Gruyter, Berlin (W), pp 179–187

Hajto T, Hostanska K (1986) An investigation of the ability of *Viscum album*-activated granulocytes to regulate natural killer cells in vivo. Clin Tri J 23:345–358

Hajto T, Lanzrein C (1986) Natural killer and antibody-dependent cellmediated cytotoxicity activities and large granular lymphocyte frequencies in *Viscum album*-treated breast cancer patients. Oncology 43:93–97

Hamprecht K, Handgretinger R, Voetsch W, Anderer FA (1987) Mediation of human NK-activity by components in extracts of *Viscum album*. Int J Immunopharmacol 9:199–209

Havas L (1937) Effects of colchicin and of *Viscum album* preparations upon germination of seeds and growth of seedlings. Nature (London) 139:371–372

Ichev K, Ovtscharoff W, Bergmann P, Franz H, Venkov L (1985) Mistletoe lectin I binding sites on the synaptosomes of the rat cerebral cortex. Acta Histochem 77:133–138

56

Jordan E (1985) Chemische und immunologische Untersuchungen von Polysacchariden und anderen hochmolekularen Inhaltsstoffen aus *Viscum album* (L.). Thesis, München

Leroi R (ed) (1987) Misteltherapie: Eine Antwort auf die Herausforderung Krebs. Die Pioniertat Rudolf Steiners und Ita Wegmans. Freies Geistesleben, Stuttgart

Luther P (1974) Ein präzipitierendes Anti-B aus *Viscum album* L. Dtsch. Gesundh. wes. 29:1487–1488

Luther P, Becker H (1986) Die Mistel: Botanik, Lektine, medizinische Anwendung. Volk und Gesundheit, Berlin

Luther P, Mehnert WH, Graffi A, Prokop O (1973) Reaktionen einiger antikörperähnlicher Substanzen aus Insekten (Protektine) und Pflanzen (Lektine) mit Ascites-Tumorzellen. Acta Biol Med Ger 31:K11–K18

Luther P, Theise H, Chatterjee B, Karduck D, Uhlenbruck G (1980) The lectin from *Viscum album* L.: isolation, characterization, properties and structure. Int J Biochem 11:429–435

Lutsch G, Noll F, Ziska P, Kindt A, Franz H (1984) Electron microscopic investigations on the structure of lectin I from *Viscum album* L. FEBS Lett 170:335–338

Lutsik MD (1975) The antitumor activity of phytohemagglutinin from *Viscum album* L. Dokl Acad Sci USSR Ser B No 6; 541–544

Metzner G, Franz H, Kindt A, Fahlbusch B, Süss J (1985) The in vitro activity of lectin I from mistletoe (ML I) and its isolated A- and B-chains on functions of macrophages and polymorphonuclear cells. Immunobiology 169:461–471

Metzner G, Franz H, Spangenberg P, Augsten K (1986) Effects of lectin I from mistletoe on platelets. In: Int Meet *Viscum album*, Heidelberg

Metzner G, Franz H, Kindt A, Schumann I, Fahlbusch B (1987) Effects of lectin I from mistletoe (ML I) and its isolated A and B chains on human mononuclear cells: mitogenic activity and lymphokine release. Pharmazie 42:337–340

Metzner G, Augsten K, Spangenberg P, Franz H (1988) In vitro effects of lectin I (ML I) from mistletoe *(Viscum album)* on human platelets: a comparison study of functional and morphological finding. (in preparation)

Neumann D, Nieden U zur, Ziska P, Franz H (1986) Are the mistletoe lectins storage proteins? In: Bøg-Hansen TC, Van Driessche E (eds) Proc 7th Int Lectin Meet, Brussels 1985, Lectins: biology, biochemistry, clinical biochemistry, vol 5. De Gruyter, Berlin (W), pp 67–73

Olsnes S, Stirpe F, Sandvig K, Pihl A (1982) Isolation and characterization of viscumin, a toxic lectin from *Viscum album* L. (mistletoe). J biol Chem 257:13263–13270

Pfüller U, Franz H, Pfüller K, Ziska P (1986) Chemical modification of the carbohydrate and tyrosyl residues of mistletoe lectin I and its lipophilization/hydrophilization. Effect on lectinological and biological behavior. 10th Arbeitstagung Lektine, Berlin

Pfüller U, Franz H, Pfüller K, Junghahn I, Bielka H (1988) Biological inactivation of mistletoe lectin I and ricin. In: Freed D, Bøg-Hansen TC (eds) Lectins: biology, biochemistry, clinical biochemistry, vol 6. Sigma Chemical Company, St. Louis, Missouri, USA, pp 299–304 (in preparation)

Portalupi E (1987) Il vischio nella terapia dei tumori. Ver Krebsforsch, Arlesheim, Schweiz

Raabe F, Storch H (1987) Untersuchungen zur Therapie des Maus-Plasmacytoms mit Mistellektin I. Wiss Z Karl-Marx-Univ Leipzig Math-Naturwiss R 36:535–543

Ribéreau-Gayon G, Jung ML, Baudino S, Sallé G, Beck JP (1986) Effects of mistletoe (*Viscum album* L.) extracts on cultured tumor cells. Experientia 42:594–599

Samtleben R, Kiefer M, Luther P (1985) Characterization of the different lectins from *Viscum album* (mistletoe) and their structural relationships with the agglutinins from *Abrus precatorius* and *Ricinus communis*. In: Bøg-Hansen TC, Breborowicz J (eds) Proc 6th Lectin Meet, Poznan 1984, Lectins: biology, biochemistry, clinical biochemistry, vol 4. De Gruyter, Berlin (W), pp 617–626

Samuelsson G (1973) Mistletoe toxins. Syst Zool 22:566–569

Sargiacomo M, Hughes RC (1982) Interaction of ricin-sensitive and ricin-resistant cell lines with other carbohydrate-binding toxins. FEBS Lett 141:14–18

Schröder G (1982) Einfluß eines Extraktes aus *Viscum album* L. auf Induktion, Wachstum- DNS- und Histongehalt von Crown-Gall-Tumoren. Thesis, Hohenheim

Schütt C, Pfüller U, Siegel E, Walzel H, Franz H (1988) Selective killing of human monocytes by immunotoxins containing partially denaturated mistletoe lectin I, vol 6. Sigma Chemical Company, St. Louis, Missouri, USA, pp 299–304 (in preparation)

Sehrt I, Luther P, Franz H, Kindt A, Samtleben R (1985) The effect of toxic lectins on the histamine release from human basophil granulocytes. In: Bøg-Hansen TC, Breborowicz J (eds) Proc 6th Lectin Meet, Poznan 1984, Lectins: biology, biochemistry, clinical biochemistry, vol 4. De Gruyter, Berlin (W), pp 53–62

Stirpe F, Legg RF, Onyon LJ, Ziska P, Franz H (1980) Inhibition of protein synthesis by a toxic lectin from Viscum album L. (mistletoe). Biochem J 190:843–845

Stirpe F, Sandvig K, Olsnes S, Phil A (1982) Action of viscumin, a toxic lectin from mistletoe, on cells in culture. J Biol Chem 257:13271–13277

Suzuki H, Franz H, Yamamoto T, Iwasaki Y, Konno H (1988) Identification of normal, microglial population in human and rodent nervous tissue using lectin-histochemistry Neuropathol Appl Neurobiol 14:221 bis 227

Taubert G, Pfüller K, Franz H, Krug H, Binder K (1988) Wachstumsverhalten des Ehrlich-Ascites-Tumors nach in vitro-Inkubation mit Mistelpräparaten. In: 9th Kongr Ges Geschwulstbekämpfung DDR, 22.–25.2., Leipzig

Vang O, Larsen K, Bøg-Hansen TC (1986) A new quantitative and highly specific assay for lectin-binding activity. In: Bøg-Hansen TC, Van Driessche E (eds) Proc 7th Int Lectin Meet, Brussels 1985, Lectins: biology, biochemistry, clinical biochemistry, vol 5. De Gruyter, Berlin (W), pp 647–644

Vester F (1977) Über die kanzerostatischen und immunogenen Eigenschaften von Mistelproteinen. Krebsgeschehen 9:106–114

Wagner H, Jordan E (1986) Nachweis und quantitative Bestimmung von Lektinen und Viscotoxinen in Mistelpräparaten. Arzneim-Forsch 36:428–433

Wagner M, Wagner B, Franz H, Ziska P (1978) Die Verwendung von Ferritinkonjugaten eines Lektins der Mistel zur elektronenmikroskopischen Lokalisation von Zelloberflächenrezeptoren. Acta Biol Med Ger 37:1537–1542

Walzel H, Mix E, Jenssen HL, Ziska P, Franz H (1982) Estimation of toxicity of the mistletoe I lectin using the footpad swelling test in mice. Acta Histochem 71:41–42

Wilson MB, Nakane PK (1978) Recent developments in the periodate method of conjugating horse radish peroxidase (HRPO) to antibodies. In: Knapp W, Holubar K, Wick G (eds) Immunofluorescence and related staining techniques, Elsevier, Amsterdam New York, pp 215–224

Yamamoto T, Franz H, Suzuki Y, Iwasaki Y (1988) Mistletoe lectin I and ricin in neurohistochemistry. In: Freed D, Bøg-Hansen TC (eds) Lectins: biology, biochemistry, clinical biochemistry, vol 6. Sigma Chemical Company, St. Louis, Missouri, USA, pp 715–719

Ziska P, Eifler R, Franz H (1979) Chemical modification studies on the D-galactopyranosyl binding lectin from the mistletoe Viscum album L. Acta Biol Med Ger 38:1361–1363

Ziska P, Franz H (1985) Determination of lectin contents in commercial mistletoe preparations for cancer therapy using the ELISA technique. In: Bøg-Hansen TC, Breborowicz J (eds) Proc 6th Int Lectin Meet, Poznan 1984, Lectins: biology, biochemistry, clinical biochemistry, vol 4. De Gruyter, Berlin (W), pp 473–480

Ziska P, Franz H, Kindt A (1978) The lectin from Viscum album L. Purification by biospecific affinity chromatography. Experientia 34:123–124

Additions in Print

Holtskog et al. (1988) described a toxic lectin in Iscador®, a mistletoe preparation widely used against cancer. This lectin is closely related to viscumin (ML I). When anti-ML I antibodies are present in the culture medium Vero cells were strongly but not completely protected against Iscador®. The investigated main cytotoxic protein in Iscador® binds to terminal nonreducing galactose residues. Addition of galactose, lactose or melibiose to the medium blocked the cytotoxic activity. The cytotoxic lectin was retained on a desialated fetuin column and on an anti-ML I-column. The authors emphasize that the molecular weight of this compound is close to, but not identical with, that of viscumin (ML I) and do not exclude that ML I (viscumin) like ricin is synthesized as one polypeptide chain which is later cleaved to yield on A and B chain. The inhibition of protein synthesis of Iscador® was compared with that of ML I. Holtskog et al. found a large difference in the sensitivity of BHK cells to Iscador® and ML I (viscumin) while the difference was smaller for the other cell lines tested.

Investigations of Hajto et al. (1988) confirmed the reviewed concept that ML I and its chains may exhibit immunomodulating effects and therefore be of importance in tumor therapy. They assume that the B chain is responsible for the stimulation of NK cells and for triggering some unspecific inflammatory events. After injection of ML I and its A and B chain small amounts of γ-interferon (IFN-γ) are released. Perhaps the B chain also enhances the production of tumor necrosis factor by murine 1774 macrophages or by human peripheral blood cells. ML II seems to exhibit immunomodulatory effects, too.

Additional experiments by Franz et al. (1989) demonstrated that the A chain of ML I releases not only IL-2 but also IL-1 from human mononuclear cells. For IL-1 estimation the thymocyte costimulator assay was used. Addition of anti IL-1 antiserum from rabbit inhibited the thymocyte-stimulating effect of the supernatant of A chain-treated human mononuclear cells.

References

Franz H, Friemel H, Buchwald S, Plantikow A, Kopp J, Körner I-J (1989) The A chain of Lectin I from European mistletoe (Viscum album) induces interleukin-1 and interleukin-2 in human mononuclear cells. In: Kocourek, J. (ed.) Lectins: biology, biochemistry, clinical biochemistry, vol 7. Sigma Chemical Company St. Louis, Missouri, USA, in press

Hajto T, Kostonka K, Vehmeyer K, Gabius H-J (1988) Immunomodulatory effects by mistletoe lectin. In: Gabius H.-J, Nagel G.A. (eds.) Lectins and Glycoconjugates in oncology. Springer, Berlin Heidelberg New York London Paris Tokyo, pp 199–206

Holtskog R, Sandvig K, Olsnes S (1988) Characterization of a toxic lectin in Iscador®, a mistletoe preparation with alleged cancerostatic properties. Oncology 45: 172–179

3 Mechanism of Action of Ricin and Related Toxic Lectins on the Inactivation of Eukaryotic Ribosomes

Yaeta Endo

3.1 Introduction

There is a group of cytotoxic proteins acting on eukaryotic ribosomes including those from plants (ricin, abrin, mistletoe lectin I and modeccin), fungi (α-sarcin) and bacteria (Shiga toxin and Vero toxin). These toxins have been known to catalytically and irreversibly inactivate 60S ribosomal subunits and in this way inhibit peptide chain elongation (Gale et al. 1981; Reisbig et al. 1981). α-Sarcin is a novel ribonuclease that hydrolyzes the phosphodiester bond on the 3′ site of the guanosine at position 4325 in 28S rRNA. This is the sole cleavage catalyzed by the toxin and this single break accounts entirely for its cytotoxicity (Schindler and Davis 1977; Endo and Wool 1982; Endo et al. 1983; Chan et al. 1983). On the other hand, the molecular mechanism of ricin however, had long been obscure. The action of ricin as a nuclease has been investigated by Mitchell et al. (1976), as a protease by Lugnier and Dirheimer (1976) and as a phosphatase or kinase by Houston (1978). These efforts have not shown a specific enzymatic activity associated with the toxin. But recently it was argued by Obrig et al. (1985) that the toxin is a ribonuclease since some toxins including ricin are able to hydrolyze naked 5S and 5.8S rRNAs.

Our initial experiment was to test whether the ribonuclease activity of the ricin A-chain, if any, is involved in the inactivation of ribosomes as e.g. α-sarcin. We determined the nucleotide sequences of 5′ and 3′ termini of each rRNA species from ricin-treated ribosomes and found that even 100 times molar excess of the A-chain over the ribosome did not hydrolyze any rRNA species, either exo- or endonucleolytically (Endo and Tsurugi 1986), which is consistent with the results of Mitchell et al. (1976). Instead, we noticed that 28S rRNA from ricin A-chain-treated ribosomes migrates more slowly during gel electrophoresis than that of the control. Further experiments revealed that ricin A-chain inactivates eukaryotic ribosomes by cleaving an N-glycosidic bond close to the α-sarcin site in 28S rRNA (Endo et al. 1987; Endo and Tsurugi 1987a). The specificity of the effect of the A-chain on ribosomes is remarkable. Not only are 5S, 5.8S and 18S rRNAs unaffected, but only one N-glycosidic bond in 28S rRNA is cleaved. The site of cleavage in rat 28S rRNA is the adenosine at position 4324 in the sequence

R S
↓ ↓

AGUACGAGAGGAAC (R = ricin site, S = α-sarcin site) which is known to be highly conserved (Chan et al. 1983). The region of the ribosome that contains the ricin- and α-sarcin-sensitive sequence must be important for the function because it is conserved and because hydrolysis of a single N-glycosidic bond or a single phosphodiester bond in that sequence inactivates the ribosomes. We have suggested that this domain in the large nucleic acid of the 60S subparticles is involved in the binding of aminoacyl-tRNA to the ribosomes (Endo and Wool 1982). The A-chain also acts on deproteinized 28S rRNA causing a cleavage at the same site as

60

that in the ribosomes however, at a reduced rate. Furthermore, the A-chain retains its activity and its specificity when the substrate is an oligomer of 553 nucleotides derived from the 3' end of 28S rRNA that has the toxin site of action (Endo and Tsurugi 1978b). Further characterization of the reaction was done by determining the enzymatic parameters of the A-chain using rat liver ribosomes and naked rRNA as substrates. It is remarkable to find that the Michaelis constant (Km) for ricin A-chain with naked 28S rRNA is the same as that for ribosome particles (Endo and Tsurugi 1987b). This suggested that the A-chain is able to bind to the same site on deproteinized 28S rRNA as in ribosomes and with a similar affinity. In contrast, the turnover number (Kcat) of the A-chain with naked 28S rRNA as substrates was much lower compared to that with ribosomes. This fact suggested that rat liver ribosomal protein(s) may condition ricin action at a step after binding (Endo and Tsurugi 1987b).

Similar activity on 28S rRNA is also exhibited by the other related proteins such as abrin, modeccin, mistletoe lectin I, Shiga toxin, the protein from wheat germ and pokeweed antiviral protein (PAP) (Endo et al. 1988a, b; Endo 1988). The evidence suggests that RNA N-glycosidase activity, originally discovered in the ricin A-chain, is a general mechanistic pathway for ribosome inactivation.

3.2 The Mechanism of Action of Ricin

3.2.1 The Effect of Ricin on 28S rRNA in Rat Liver Ribosomes

When rat liver ribosomes were treated with ricin A-chain at a molar ratio of 1:960 (ricin:ribosomes), the activity of the ribosomes was decreased by more than 85% in poly(U)-directed polyphenylalanine synthesis (data not shown), indicating that the toxin catalytically inactivates ribosomes. To gather evidence that RNA is the toxin target, total rRNAs were extracted from the ricin-treated ribosomes and analyzed by composite-gel electrophoresis. As shown in Figure 3.1A, comparison of electrophoretic mobility of treated (lane 2) vs untreated (lane 1) rRNAs showed identical mobility of 18S rRNA, while a mobility difference between the two 28S rRNAs

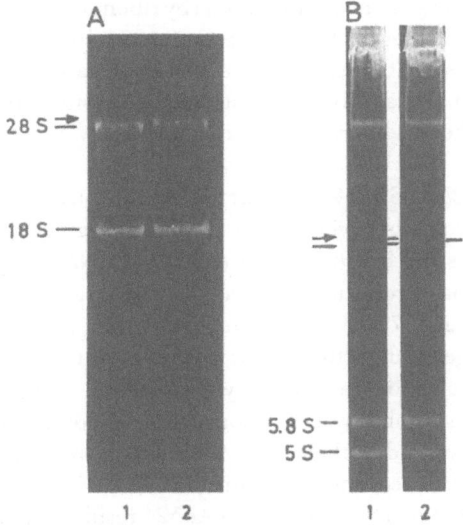

Fig. 3.1 Analysis by gel electrophoresis of RNAs from ricin A-chain treated rat liver ribosomes. Rat liver ribosomes (3.49×10^{-7} M) were incubated with ricin A-chain (3.12×10^{-10} M) at 37 °C for 10 min and the RNA was extracted with sodium dodecyl sulfate and phenol (Endo et al. 1987)

A: 2 µg of RNAs from control (lane 1) and ricin-treated ribosomes (lane 2) were analyzed by 2.5% acrylamide- 0.5% agarose composite-gel system

B: 20 µg of RNAs from control (lane 1), and ricin-treated ribosomes (lane 2) were analyzed by 3.5% polyacrylamide gel electrophoresis. RNA bands were visualized with ethidium bromide.

Arrow denotes 28S rRNA (A) or the fragment (B) altered in its migration rate by ricin.

was clearly visible, as indicated by the arrow. This small but definitive difference between the two 28S rRNAs was reproducible from experiment to experiment. This finding suggested that the ricin A-chain modifies 28S rRNA, and in some way, results in slow migration on the gel. That ricin does not act as an endoribonuclease or as an exoribonuclease has been clearly demonstrated (Endo and Tsurugi 1986). The observed resistance of RNA cleavage by ricin was in disagreement with previously reported data (Obrig 1984).

3.2.2   The Site of Modification in 28S rRNA

To identify the site of modification on the 28S rRNA, we searched for modified rRNA fragments among the many RNA fragments that are normally generated by the contaminated ribonuclease(s) which associates ribosomes during their preparation. A fragment of about 550 nucleosides clearly had slower mobility compared to ricin-untreated ribosomes, as shown by an arrow in lane 2 of Figure 3.18. This difference in the rate of migration disappeared when both treated and untreated samples were analyzed by gel electrophoresis in the presence of 7 M urea (data not shown). This observation suggested that the mobility shift of the ricin-treated rRNA was due to the possible change in the conformational or chemical modification rather than change in length.

In order to determine the origin of the modified fragment, we isolated the fragment and determined the nucleotide sequences of both 5′- and 3′-terminal regions (data not shown). Both modified and unmodified 550 nucleotide fragments revealed identical 5′- and 3′-end sequences by an enzymatic digestion method and we localized this fragment to the 3′-terminal 553 nucleotides of the 28S rRNA (Endo et al. 1987). The occurrence of this fragment in the ribosomal preparation has been reported previously by Choi (1985). This result indicated that the modified region may be located in the middle portion of the modified fragment. Further nucleotide sequence analysis revealed the absence of bands corresponding to G4323 and A4324 in the modified fragment, whereas these bands were present in the unmodified fragments as shown by arrows in Figure 3.2A.

This striking observation immediately suggested that these two nucleotides have been modified by an unknown enzymatic activity of the ricin A-chain. A possibility of a known type of modification of the residues can be excluded because of its resistance to hydrolysis by ribonuclease T2 which is known to recognize most kinds of modified bases (Uchida and Egami 1971; Fig. 3.2A, lanes 1 and 2). The higher susceptibility to hydrolysis of C4322 by B. cereus ribonuclease of the modified over the unmodified fragment (see Fig. 3.2A, lane 10), suggests that as a result of the modification of either or both G4323 and A4324, the neighboring C residue has been widely exposed to digestion by the ribonucleases.

Since it is possible that modification of either G4323 or G4324 results in the ribonuclease resistance of both nucleotides, we examined the cleavage behavior of the modified fragment by chemical methods. It is known that various amines and hydroxide ions cleave the RNA strand by a β-elimination reaction if the base of the nucleoside residue is removed leaving an aldehyde radical at C1 of the ribose (Kochetov and Budovskii, 1972). As shown in Figure 3.2A, lanes 13 and 14, partial alkaline hydrolysis shows stronger radioactive bands corresponding to the G4323 and A4324 residues, compared to other nucleotides (as shown by arrowheads). This result suggested that the modification imparts increased lability of the phosphodiester bonds surrounding A4324, since the 5′ end of the fragment was radio labeled. Treatment of the modified and unmodified fragments with aniline at acidic pH, according to Peattie (1979), also resulted in chain scission at positions apparently corresponding to the G4323 and A4324 residues. This

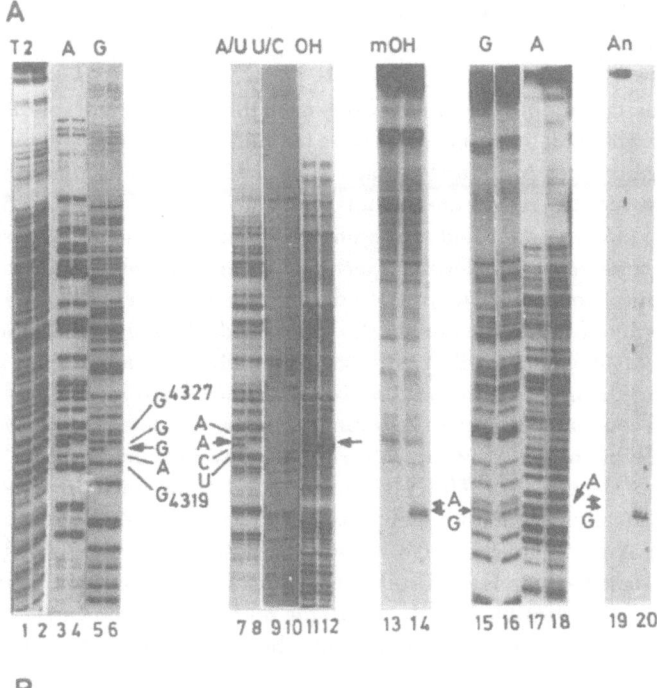

A

| T2 | A | G | | A/U | U/C | OH | | mOH | | G | A | | An |

G4327
G   A
G   A
A   C
G4319 U

1 2 3 4 5 6   7 8 9 10 11 12   13 14   15 16 17 18   19 20

B

ricin site
↓
4315 CUCAGUACGAGAGGAACCGCA 4335
↑
α-sarcin site

Fig. 3.2 Radioautographs of sequencing gels of the modified region of the 3'-terminal fragment of 28S rRNA (Endo et al. 1987)

A: The fragments of 553 bases from control and ricin-treated ribosomes were made radioactive at the 5' terminus with (γ-^{32}P) ATP using polynucleotide kinase and were partially digested with ribonuclease T1 (G), ribonuclease U2 (A), ribonuclease Phy M (A/U), ribonuclease T2 (A/U/G/C), or ribonuclease from *B. cereus* (U/C). The partial alkaline digests obtained under mild (1 min) and ordinary (15 min) conditions were designated mOH and OH respectively. The aniline-induced products were designated An. The digests of the fragments from control (odd-numbered lanes) and ricin-treated ribosomes (even-numbered lanes) were separated by 10% acrylamide gel electrophoresis at 1.2 kV for either 4 h (lanes 1–12) or 6 h (lanes 13–20). The arrows denote the bases missing in ricin-treated RNA, corresponding to G4323 and A4324. The arrowheads denote the bands developed by alkaline (lane 14) and aniline treatment (lane 20) which also apparently correspond to G4323 and A4324.
B: Nucleotides in 28S rRNA are numbered from its 5'-terminal and according to Chan et al. (1983).

result also indicated that the fragment was sensitive at both phosphodiester bonds surrounding the A4324 residue (Fig. 3.2B). If the G4323 residue is also modified, the band of C4322 will appear on the gel with this treatment. These results suggested that the base of A4324 is cleaved, leaving an aldehyde radical at C1 of the ribose rather than showing that the bases of both residues are missing or are severely damaged (Endo et al. 1987).

3.2.3 An RNA N-Glycosidase Activity of the Ricin A-Chain

Bases were recovered by ion-exchange column chromatography from the 50 % ethanol soluble fraction of the reaction mixture in which rat liver ribosomes were treated with the ricin A-chain. The fraction from ricin-treated ribosomes contained a major spot of UV absorbing material which is absent in the fraction from control ribosomes by thin-layer chromatography (Fig. 3.3, lanes 5 and 6). The material of the spot was identified as adenine (Ade) from its R_f value. It must be noted that although there are some minor nucleoside spots on the thin-layer plate, none of them correspond to guanine (Gua). Then, the amount of adenine was quantitated by densitometry using pure adenine as standards and the molar ratio of released adenine to ribosome was calculated (Table 3.1). The result indicates that every ribosome liberated adenine in a nearly stoichiometric ratio (0.78 to 0.84 mol adenine per 1 mol of ribosome). Therefore, considering the results shown above, we concluded that ricin A-chain inactivates ribosomes by cleaving the N-glycosidic bond at A4324, but not at G4323, in 28S rRNA (Endo and Tsurugi 1987).

We next studied the mechanism of the reaction. The N-glycosidic bond of the nucleoside residue in 28S rRNA can be enzymatically cleaved by either phosphorolysis or hydrolysis. Olsnes et al. (1975) observed that ricin A-chain inactivated ribosomes simply in Tris/KCl/MgCl₂ medium and hence claimed that ricin A-chain is a hydrolytic enzyme. However, this point has not been formally established since it is possible that the ribosome preparation even carries a trace of phosphate. In the phosphorolytic mechanism, phosphate should be incorporated into

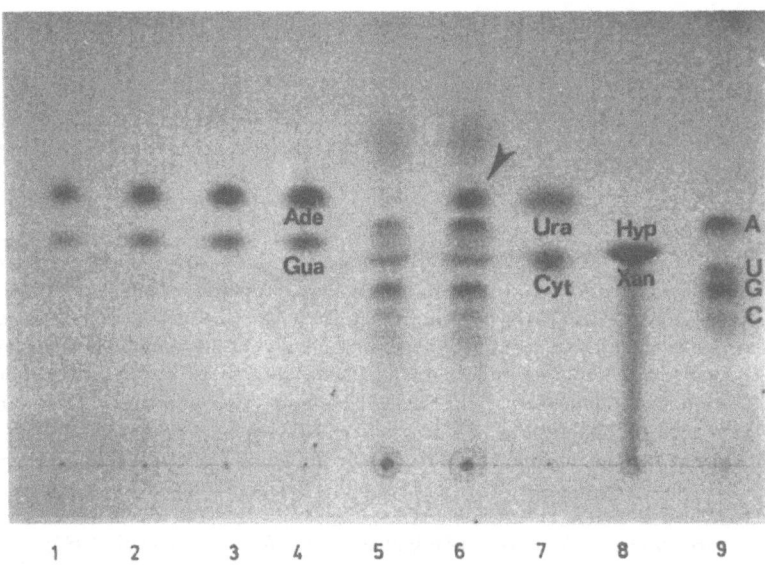

Fig. 3.3 Identification of a base liberated from the ribosomes by the action of ricin A-chain. Bases were isolated from the reaction mixture and separated on a silica gel plate with 1-butanol/methanol/water/ammonia (60/20/20/1) as a solvent (Endo and Tsurugi 1987a)
Lanes 1–4: standards of adenine and guanine containing various amounts
lane 5: bases from control ribosomes
lane 6: bases from ricin A-chain-treated ribosomes
lanes 7–9: authentic markers as shown (Hyp represents hypoxanthine).
Arrow head represents the spot released from the ribosomes by the action of the ricin A-chain.

Table 3.1 Quantitation of adenine released from the ribosomes by the action of the ricin A-chain. Details are described in Experimental Procedures

Experiment	Ribosome[1] [nmol]	Recovery[2] [%]		Amount of base[3] measured [nmol]		Total base released [nmol]		Molar ratio[4]	
		Ade[5]	Gua[6]	Ade	Gua	Ade	Gua	Ade	Gua
Untreated Ricin A-chain treated	2.33	73.6	67.1	ND[7]	ND	–	–	–	–
1	2.33	72.7	66.3	1.43	ND	1.97	–	0.84	–
2	2.33	71.1	66.5	1.34	ND	1.89	–	0.81	–
3	2.33	75.9	69.2	1.38	ND	1.82	–	0.78	–

[1] The calculations are based on a value vor $\Sigma^{1\%}_{260\,nm}$ of 100 and a molecular weight of 4.3×10^6.
[2] calculated from the recovery of [³H] adenine and [¹⁴C] guanine
[3] estimated by measuring the relative intensity to the standard spots
[4] released moles of the base per mole of ribosome
[5] adenine
[6] guanine
[7] not detectable

Table 3.2 Search for incorporation of phosphate into 28S rRNA during cleavage of N-glycosidic bond of A4324[1]

Reaction conditions	Reaction	pmol in 100 µl of reaction mixture
1.0 ng ricin A-chain and ribosomes, 37°C, 10 min	aniline-sensitive rRNA	34.9
	phosphate incorporated	0
10.0 ng ricin A-chain and ribosomes, 37°C, 10 min	aniline-sensitive rRNA	34.9
	phosphate incorporated	0.2
10.0 µg ricin A-chain and naked rRNA, 37°C, 60 min	aniline-sensitive rRNA	22.3
	phosphate incorporated	0.1

[1] Ribosomes or naked total rRNA (34.9 pmoles) were incubated with ricin A-chain in the presence of [³²P]i in 100 µl buffer (25 mM Tris-HCL, pH 7.6, 25 mM KCL and 5 mM MgCl₂). The treatment of ribosomes or naked rRNA with ricin A-chain resulted in 100 and 64 % cleavage of the N-glycosidic bond of A4324 respectively. The moles of phosphate incorporated were represented after subtraction of those of the toxin-untreated ribosomes (non-specific adsorption) which were usually between 2.3 and 3.7 pmol per 34.9 pmol 28S rRNA.

the nucleside residue, forming ribose-1-phosphate. To test this possibility, ribosomes were treated with ricin A-chain in the presence of [³²P]-phosphate and the incorporation of the radioactivity into the 28S rRNA fraction was measured (Table 3.2). There was little incorporation of phosphate into 28S rRNA as it was calculated that less than 1 mol of phosphate/100 mol of modified 28S rRNA was incorporated. A possibility that the specific radioactivity was diluted by free phosphate associated with ribosomes was ruled out because essentially the same result

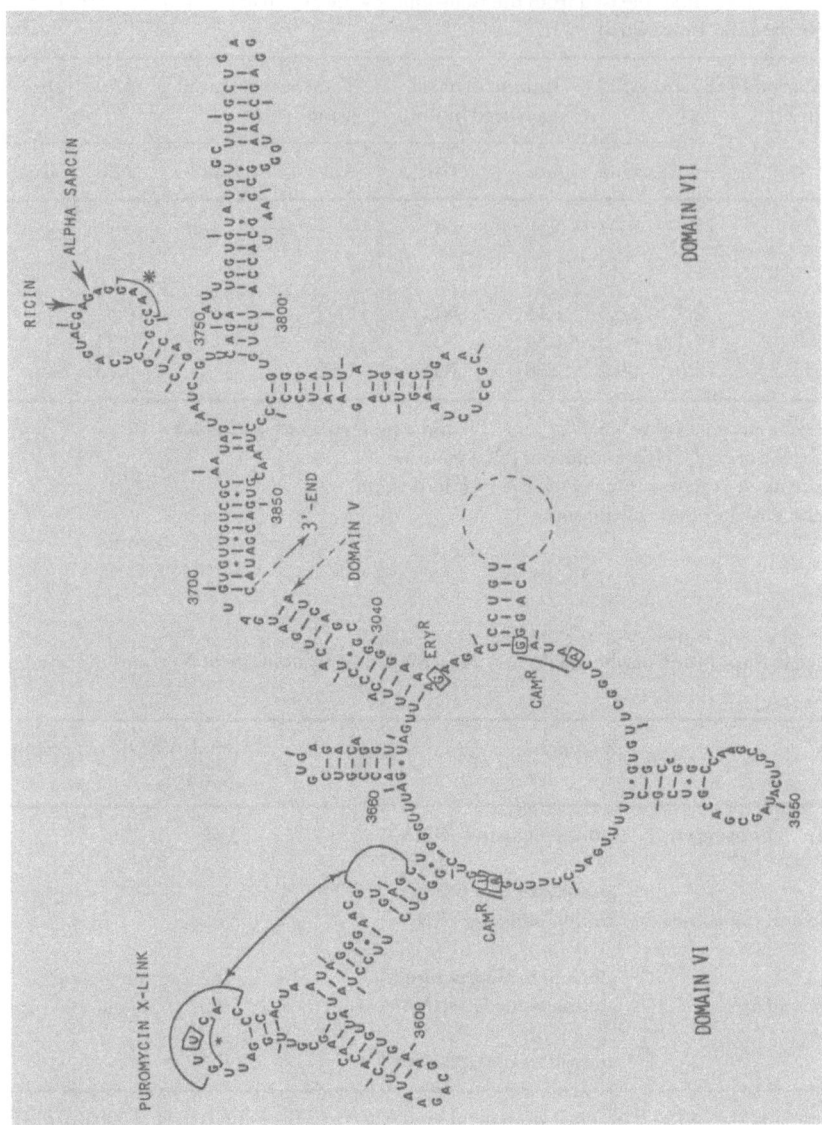

Fig. 3.4 Alignment of the ricin and α-sarcin site in the proposed secondary structure of 28S rRNA of *Xenopus laevis*

The structure of part of domain VI and domain VII of *Xenopus* 28S rRNA is taken from Clark et al. (1984) and Endo (1988). The bases of mutation lead to erythromycin resistance or chloramphenicol resistance in *E. coli* and are denoted as ERY® and CAM®, respectively. Puromycin X-link represents the site where puromycin cross-links to *E. coli* 23S rRNA. Large and small asterisks denote the tetranucleotide GAAC and GUUC respectively. The sites of action of ricin A-chain and α-sarcin on *Xenopus* 28S rRNA in the ribosomes have been confirmed experimentally (Endo unpublished).

66

was obtained when the naked rRNA was incubated (see below), which is expected to carry less phosphate than ribosomes (lower panel). Thus, it is concluded that the ricin A-chain does not act as a phosphorolytic enzyme (Endo and Tsurugi 1987).

So far we have shown that ricin A-chain inactivates eukaryotic ribosomes by cleaving the N-glycosidic bond of A4324 in 28S rRNA probably in a hydrolytic fashion. N-glycosidase (EC 3.2.2) is a class of enzyme that cleaves N-glycosidic bonds in a hydrolytic fashion, and within this group enzymes are found acting on such diverse substrates as uridine (Magni et al. 1975), NAD (Kaplan et al. 1951) and S-adenosylhomocysteine (Duerre 1962). A different group of N-glycosidases, which hydrolyze base-sugar bonds in DNA, were recently discovered in bacteria (Lindahl 1976; Lindahl et al. 1977). They specifically attack DNA containing damaged or non-conventional bases and are believed to function in DNA repair. However, ricin A-chain as described is totally different from them in the respect that the toxin cleaves only one particular N-glycosidic bond of approximately 7,000 present in eukaryotic rRNA.

What is now clear is that ricin A-chain inactivates eukaryotic ribosomes by hydrolyzing only one particular N-glycosidic bond of eukaryotic rRNA. The question, then, is what is the functional correlate of the structure at the ricin site in the ribosomes. The site of action of the ricin A-

R S
↓ ↓

chain in 28S rRNA is on A4324 in the sequence AGUACGAGAGGAAC, close to the α-sarcin site (R = ricin site, S = α-sarcin site). This sequence is highly conserved. It occurs in eukaryotes, in yeast 25S rDNA (Veldman et al. 1981) and in rat liver 28S rRNA (Endo and Wool 1982; Chan 1983); it also occurs in prokaryotes, in *Escherichia coli* 23S rDNA (Noller 1984). The region of the ribosomes that contains a ricin and α-sarcin sensitive sequence must be important for the function because it is conserved and because hydrolysis of a single N-glycosidic bond or the hydrolysis of a single phosphodiester bond in that sequence inactivates the ribosomes. We have suggested that this region is involved in EF-1-catalyzed binding of aminoacyl-tRNA to the ribosomes. This is known to be the reaction that is inhibited by the toxins. Additional evidence came from the search for the sequence of the toxin site for nucleotides complementary to invariant or semiinvariant bases in tRNA, and we discovered a potential interaction (Endo and Wool 1982). The complementarity is between the tetranucleotide GAAC in that sequence (Fig. 3.4, large asterisk) and the invariant sequence GTΨC in loop IV of all eukaryotic tRNAs except the initiator tRNA. In addition, it is interesting to find that there is an another potential complementarity between GAAC near the toxin sites and GUUC in the puromycin reactive site (small asterisk). However, the physiological importance of these interactions remains to be established. Another possibility is that peptidyl-tRNA actually binds at a distant site but in some way alters the structure of that region, which then might be involved in some other reaction essential for protein synthesis. It therefore remains to be established for the functional roles of that domain.

3.2.4   The Characteristics of the Enzymatic Activity of Ricin A-Chain with Ribosomes and Naked rRNA as Substrates

Ricin treatment of rat 80S ribosomal particles, and the subsequent scission of the RNA with aniline, generated a specific fragment (the R-fragment which is derived from the 3′ side of 28S rRNA having 449 nucleotides; Fig. 3.5A), confirming the above results. A concentration of the

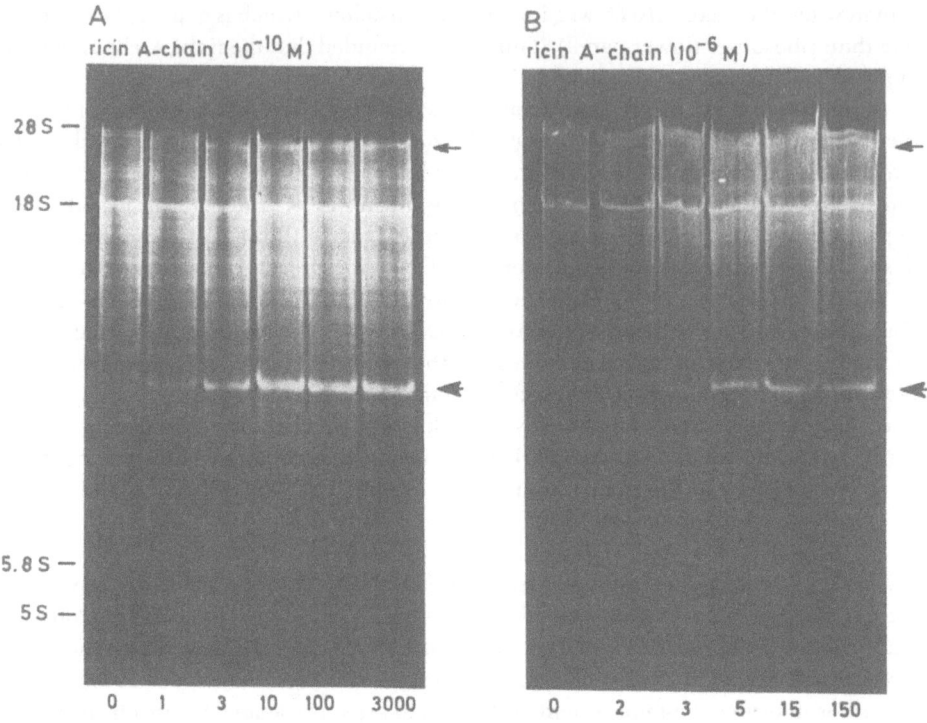

Fig. 3.5 Effect of ricin A-chain on rat ribosomes and rRNA

In A rat liver 80S ribosomes (38 µg) in 25 µl buffer A (25 mM Tris/HCl, pH 7.6; 25 mM KCl; 5 mM MgCl$_2$) and in B naked rRNA (21.9 µg) in 25 µl buffer B (25 mM Tris/HCl, pH 7.6; 100 mM NaCl, 10 mM MgCl$_2$) were incubated at 37°C with ricin A-chain in the concentrations indicated. After 10 min for ribosomes or 60 min for naked rRNA, the nucleic acids were extracted from the reaction mixture with phenol and sodium dodecyl sulfate. The RNA (5 µg) was treated with aniline at acidic pH at 60°C for 10 min to induce scission of the phosphodiester bond at the apurine site. The samples were analyzed by electrophoresis in 2.5 % polyacrylamide-0.5 % agarose composite gels. The arrowheads designate the R-fragment (Endo and Tsurugi 1987b).

A-chain of 1×10^{-10} M (an enzyme to substrate ratio of 1118) was effective in cleaving the N-glycosidic bond at A4324 in 28S rRNA and even if the concentration was increased by three orders of magnitude only the R-fragment was produced. Production of the R-fragment was complete in 10 min (Fig. 3.6C) when the concentration of ricin A-chain was 1×10^{-9} M (Fig. 3.6A). Treatment of protein free rat rRNA with ricin A-chain caused the cleavage of a N-glycosidic bond, probably at the same site in 28S rRNA as in ribosomal particles since the mobility of the fragment formed is similar (Fig. 3.5B). The naked rRNA, however, required four orders of magnitude more of the A-chain and longer incubation with the toxin as well (Fig. 3.6B, C). The result indicated that the protein acts directly on naked rRNA. Whether a smaller RNA fragment containing the ricin site is also a substrate for the A-chain was next tested. For this purpose, the 3′-terminal fragment of 28S rRNA having 553 nucleotides was isolated from untreated ribosomes and its 5′ terminus was made radioactive. The fragment was incubated with ricin A-chain and the RNA was analyzed by gel electrophoresis after treatment with aniline. The A-chain acts on this smaller fragment as is evidenced by the production of a 5′-terminal oligonucleotide having a size similar to that of the R-fragment marker (Fig. 3.7A, lanes 1–5). During the

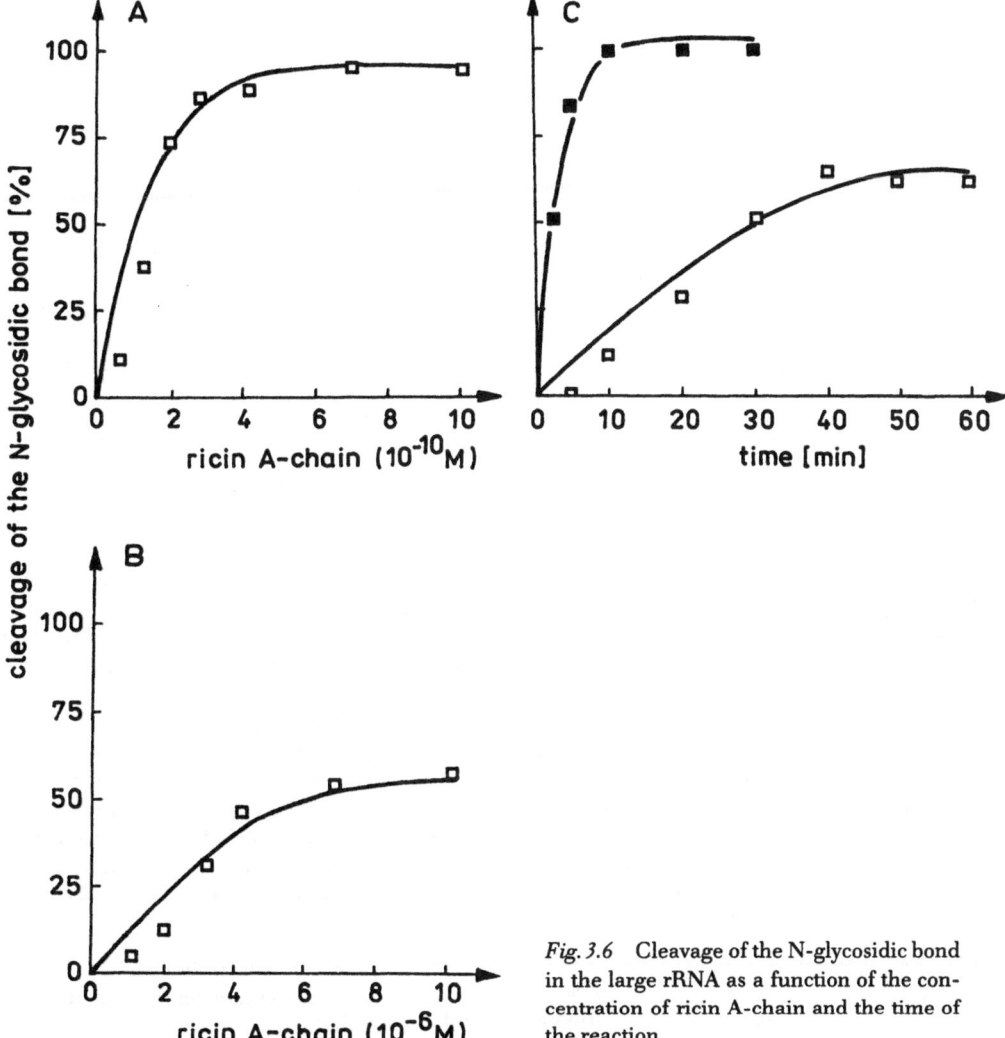

Fig. 3.6 Cleavage of the N-glycosidic bond in the large rRNA as a function of the concentration of ricin A-chain and the time of the reaction

In A uniformly [^{32}P]-labeled 80S ribosomes (114 µg; 1.5×10^8 cpm/mg RNA) were incubated for 10 min at 37 °C in 25 µl buffer A with the concentration of ricin A-chain indicated. The RNA was extracted and separated on composite gels after chain scission with aniline. The radioactivity in the R-fragment and in the 5.8S rRNA bands was determined. The extent of cleavage of the N-glycosidic bond was calculated from the molar ratio of 5.8S rRNA to the R-fragment. In B, [^{32}P]-labeled total rRNA extracted from the ribosomes was incubated for 30 min in 25 µl buffer B. In C, each substrate was incubated for the indicated period of time with ricin A-chain (1×10^{-9} M for ribosomes ■ and 5×10^{-6} for the naked ones □). The extent of cleavage of the band at each time interval was determined (Endo and Tsurugi 1987b).

course of these experiments, we found that only the magnesium ion was essential for the activity of ricin. The optimum concentration of magnesium is about 10 mM. The 5'-[^{32}P]-labeled 553-fragment that had been treated with ricin A-chain was subjected to an RNA-sequencing reaction and a pattern of bands (Fig. 3.7B, lanes 3 and 6) was observed, similar to those obtained with reference rRNA prepared from ricin-treated ribosomes (lanes 4 and 7). Treatment of the

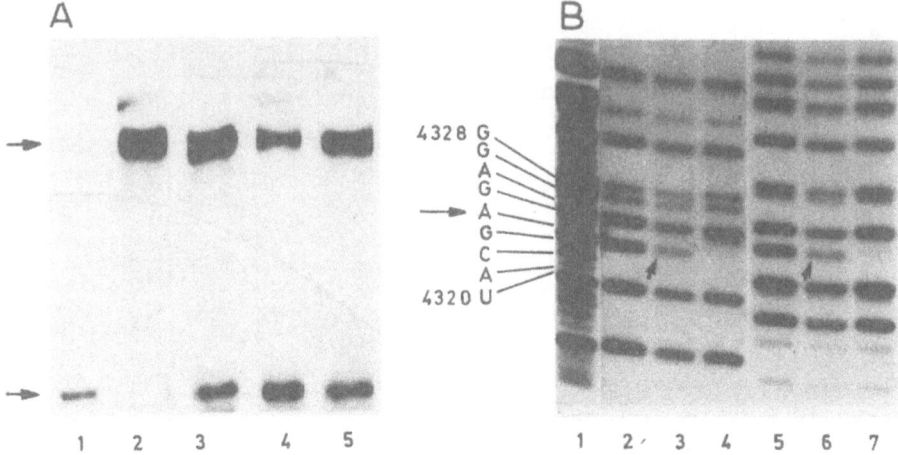

Fig. 3.7 Effect of ricin A-chain on the 553-fragment and the identification of the site of action. The 553-fragment was isolated from rat liver rRNA preparations and made radioactive with [γ-^{32}P]-ATP and T4-kinase was repurified by electrophoresis in 3.5 % polyacrylamide gels

In A the fragment (0.15 μg; 2×10^{10} cpm·mg^{-1}) was incubated in 25 μl buffer containing varying concentrations of MgCl$_2$ at 37°C for 60 min. The RNA was extracted and treated with aniline and then separated by electrophoresis in 3.5 % polyacrylamide gel. In lane 1, the 553-fragment isolated from ricin-treated rat liver ribosomes serves as a marker for the 5' fragment produced by aniline treatment (arrowhead). The 553-fragment that had been incubated without (lane 2) or with the A-chain in the presence of different concentrations of MgCl$_2$; 0.2 mM (lane 3), 10 mM (lane 4), and 20mM (lane 5), was analyzed. The arrow denotes the position of the 553-fragment on the gel. In B, portions of the [^{32}P]-labeled 553-fragment which had been treated with ricin A-chain in the presence of 10 mM MgCl$_2$ were partially digested with ribonuclease T1 (G) or ribonuclease PhyM (A/U). The digests were separated by electrophoresis in 10 % polyacrylamide gel (lanes 3 and 6). For comparison, similar experiments were carried out on the 553-fragment from non-treated ribosomes (lanes 2 and 5) and from ricin-treated ribosomes (lanes 4 and 7). The alkaline digests (lane 1) were prepared with the control 553-fragment. The arrowheads denote the ricin-site, A4324 in 28S rRNA (Endo and Tsurugi 1987b).

Table 3.3 Michaelis constant (Km) and turnover number (Kcat) of ricin A-chain[1]

Substrate	Km [μM]	Kcat [ribosomes/min]
Rat liver 80S ribosome	2.6	1777
Rat liver naked 28S rRNA	5.8	0.02

[1] The initial reaction velocity was measured using [^{32}P]-labeled ribosomes or ribosomal RNA. The parameters were calculated from the reciprocal plot of the data.

Ricin A-chain catalyzed reaction:

$$S + R + \underset{K2}{\overset{K1}{\rightleftarrows}} E\text{-}S \overset{K3}{\rightarrow} P + R$$

S: substrate, R: ricin A-chain, E-S: binary complex as an intermediate of the reaction, P: product, K: velocity constant

70

553-fragment with A-chain made the phosphodiester bonds at G4323 and at A4324 resistant to the action of ribonucleases (cf. arrowheads in Fig. 3.7).

The faint bands seen at these two sites are likely to be derived from hydrolysis in unaffected fragments. Thus, the A-chain acts on the 553-fragment to cleave the N-glycosidic bond at the same site as in ribosomes. All of the above results suggested that the A-chain recognizes a specific structure in the RNA. In addition, it is apparent that removal of ribosomal proteins decreases the sensitivity of this bond to the toxin. Thus, r-protein(s) modulate the response to the protein.

To further characterize the reaction, we determined the enzymatic parameters of the toxin (Endo and Tsurugi 1988). Preparations of [^{32}P]-labeled 80S ribosomes or of rRNA, at concentrations of 0.5 to 2.7 µM, were treated with ricin A-chain, followed by aniline, and the radioactivity in the R-fragment was determined. These steady state kinetic experiments were performed so as the initial velocity of the reaction that was assessed. From a double-reciprocal plot of the data we calculated that the apparent Michaelis constant (Km) was 2.6 µM and the turnover number (Kcat) was 1777 min^{-1} (Table 3.3). These values are in reasonable agreement with those reported by Olsnes et al. (1975) who used an indirect method. The Km which we obtained accounts for the extreme toxicity of ricin since the intracellular concentration of ribosomes is of the order of 1 µM (Blobel and Potter 1967). A similar experiment was carried out with naked rRNA and for the most part the same Km value was obtained, however, the Kcat was much lower (see Table 3.3). It should be noted that the Kcat value with naked 28S rRNA is only an approximation since these experiments were done under non-catalytic conditions. Assuming that K1 and K2 are K3 (in the reaction illustrated in the footnote to Table 3.3), the finding that there is no significant difference in the Km values with the two substrates suggests that ricin A-chain recognizes a site in naked 28S rRNA and binds there with the same affinity as in ribosomes. This interpretation conformed to the observation of Hedblom et al. (1976) who estimated the dissociation constant (Kd) for the A-chain on rat liver ribosomes, employing Scatchard plot analysis, to be 3 µM which is close to our Km value. Thus, the ricin A-chain appears to recognize a specific structure in rRNA and the recognition does not seem to require protein. Moreover, intact 28S rRNA is not essential since the 553-nucleotide fragment that contains the ricin site can serve as a substrate for the toxin. Rat liver r-protein(s) may condition ricin action at a step after binding since their removal results in a large reduction in the Kcat.

3.3   The Mechanism of Action of Ricin-Related Lectin Toxins and other Ribosome-inactivating Proteins

Proteins that catalytically inactivate eukaryotic ribosomes have been identified in the extracts of a wide variety of plants (Jimenez and Vazquez 1985; Stirpe and Barbieri 1986). These proteins can be divided into two classes. One class comprises proteins that consist of two non-identical subunits (A- and B-chain) that are joined by a disulfide bond (Gale et al. 1981). Ricin, abrin, mistletoe lectin I, modeccin and Shiga toxin are the best-studied examples of this class (Gale et al. 1981; Reisbig et al. 1981). The B-chains of these proteins bind to receptors on the surface of cells and promote the uptake of their A-chain into cells. Entry of the A-chain into the cytoplasm of a cell then results in the death of the cell by inactivation of its ribosomes. Thus, members of this class of proteins are potent inhibitors of protein synthesis in intact cells as well as in cell-free systems, and are highly cytotoxic for animals. The second class of ribosome-inactivating proteins consists of proteins that have single polypeptide chains that are also potent inhibitors of protein synthesis in a cell-free system. However, they are relatively non-toxic to intact cells or

71

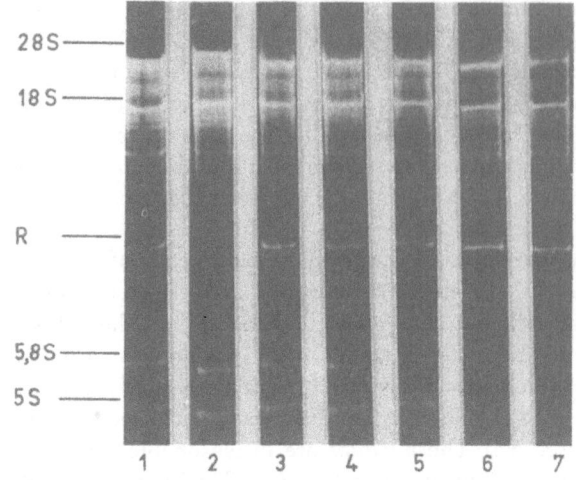

Fig. 3.8 The other ricin-related toxins and ribosome-inactivating proteins have probably the same mechanism of action on rat liver ribosomes as ricin. Rat liver ribosomes were treated with a catalytic amount of one of the following six proteins and were analyzed in the same way as in Figure 3.5 (Endo 1988)

Lane 2: control rRNA treated with aniline, lanes 1, 3–7: aniline-treated RNA from ribosomes treated with ricin as a marker (1), abrin (3), modeccin (4), PAP (5), the protein from wheat germ (6) and Shiga toxin (7).

R shows the R-fragment.

animals (Gale et al. 1981). Following the identification of pokeweed antiviral protein (PAP) as a protein that inactivates ribosomes (Irvin 1975), some 30 different single-chain ribosome-inactivating proteins have been purified from 16 different plants, and it may be that these proteins are ubiquitous in plants (Gasperi-Campani et al. 1985).

In order to determine whether any of these proteins, other than ricin, has a similar mechanism of action on ribosome inactivation, we analyzed rRNA by gel electrophoresis after treatment with aniline (see Fig. 3.5). Figure 3.8 clearly shows that treatment of rat liver ribosomes with one of the following proteins, abrin, modeccin, Shiga toxin (the former group of proteins) or PAP, the protein from wheat germ (the latter group), produced the same pattern on the gel. The production of the R-fragment was at the cost of disappearance of 28S rRNA. The results demonstrated that all of the ribosome-inactivating proteins examined so far have a similar mechanism of action on eukaryotic ribosomes as ricin.

Thus, the RNA N-glycosidase activity discovered in ricin A-chain seems to be one of the enzymatic activities generally found among the proteins that inactivate eukaryotic ribosomes.

3.4 References

Blobel G, Potter VR (1967) Studies on free and membrane-bound ribosomes in rat liver. J Mol Biol 26:279–292

Chan Y-L, Endo Y, Wool IG (1983) The sequence of the nucleotides at the α-sarcin cleavage site in rat 28S ribosomal ribonucleic acid. J Biol Chem 258:12768–12770

Choi CC (1985) Structural organization of ribosomal RNAs from Novikoff hepatoma. J Biol Chem 260:12773–12779

Duerre JA (1962) A hydrolytic nucleosidase activity on S-adenosyl-homocysteine and on 5'-methyl-thioadenosine. J Biol Chem 237:3737–3741

Endo Y (1988) Mechanism of action of ricin and related toxins on the inactivation of eukaryotic ribosomes. In: Frankel AE (ed) Immunotoxins. Nijhoff, Boston, pp 75–89

Endo Y, Tsurugi K (1986) Mechanism of action of ricin and related lectins on eukaryotic ribosomes. Nucleic Acids Res Symp Ser No 17:187–190

Endo Y, Tsurugi K (1987a) RNA N-glycosidase activity of ricin A-chain. J. Biol Chem 262:8128–8130

Endo Y, Tsurugi K (1987b) The RNA N-glycosidase activity of ricin A-chain. (Submitted)

Endo Y, Wool IG (1982) The site of action of α-sarcin on eukaryotic ribosomes. J Biol Chem 257:9054–9060

Endo Y, Huber PW, Wool IG (1983) The ribonuclease activity of the cytotoxin α-sarcin. J Biol Chem 258:2662–2667

Endo Y, Mitsui K, Motizuki M, Tsurugi K (1987) The mechanism of acton of ricin and related toxic lectins on eukaryotic ribosomes. J Biol Chem 262:5908–5912

Endo Y, Tsurugi K, Yutsudo T, Takeda Y, Ogasawara T, Igarashi Y (1988a) Site of action of a Vero toxin (VT2) from *Escherichia coli* 0157:H7 and of Shiga toxin on eukaryotic ribosomes. Eur J Biochem (in press)

Endo Y, Tsurugi K, Franz H (1988b) The site of action of mistletoe lectin A-chain on eukaryotic ribosomes. The RNA N-glycosidase activity of the protein. FEBS letters 230:

Gale EF, Cundliffe E, Reynolds PE, Richmond MH, Waring MJ (1981) In: The molecular basis of antibiotic action, 2nd edn. Wiley, New York, pp 402–547

Gasperi-Campani A, Barbieri L, Battelli MG, Stirpe F (1985) On the distribution of ribosome-inactivating proteins amongst plants. J Nat Prod 48:446–454

Hedblom ML, Cawley DB, Houston LL (1976) The specific binding of ricin and its polypeptide chain to rat liver ribosomes and ribosomal subunits. Arch Biochem Biophys 177:46–55

Houston LL (1978) Effect of the toxic caster bean protein, ricin, on the phosphorylation pattern of [^{32}P]-labeled ribosomes from mouse L cells. Biochem Biophys Res. Commun 85:131–139

Irvin JD (1975) Purification and partial characterization of the antiviral protein from *Phytolacca americana* which inhibits eukaryotic protein synthesis. Arch Biochem Biophys 169:522–528

Jimenez A, Vazquez D (1985) Plant and fungal protein and glycoprotein toxins inhibiting eukaryotic protein synthesis. Annu Rev Microbiol 39:649–672

Kaplan NO, Colowick SP, Nason A (1951) Neurospora diphosphopyridine nucleotidase. J Biol Chem 191:473–483

Kochetov NK, Budovskii EI (1972) In: Organic chemistry of nucleic acids. Plenum, New York London, pp 269–618

Lindahl T (1976) New class of enzymes acting on damaged DNA. Nature (London) 259:64–66

Lindahl T, Ljungquist S, Siegert W, Nyberg B, Sperens B (1977) DNA N-glycosidases. J Biol Chem 252:3286–3294

Lugnier AAJ, Dirheimer G (1976) Action of ricin from *Ricinus communis* L. seeds on eukaryotic ribosomal proteins. FEBS Lett. 67:343–347

Magni G, Fioretti E, Ipata PL, Natalini P (1975) Bakers yeast uridinenucleotidase purification, composition, and physical and enzymactic properties. J Biol Chem 250:9–13

Mitchell SJ, Hedblom M, Cawley D, Houston LL (1976) Ricin does not act as an endonuclease on L cell polysomal RNA. Biochem Biophys Res Commun 68:763–769

Noller HF (1984) Structure of ribosomal RNA. Annu Rev Biochem 53:119–162

Obrig TG, Moran TP, Colinas RJ (1985) Ribonuclease activity associated with the 60S ribosome-inactivating proteins ricin A, phytolaccin and Shiga toxin. Biochem Biophys Res Commun 130:879–884

Olsnes S, Fernandez-Puentes C, Carrasco L, Vasquez D (1975) Ribosome inactivation by the toxic lectins abrin and ricin. Eur J Biochem 60:281–288

Peattie DA (1979) Direct chemical method for sequencing RNA. Proc Natl Acad Sci USA 76:1760–1764

Reisbig R, Olsnes S, Eiklid K (1981) The cytotoxic activity of *Shigella* toxin. J Biol Chem 256:8739–8744

Schindler DG, Davies JE (1977) Specific cleavage of ribosomal RNA caused by alpha sarcin. Nucleic Acids Res 4:1097–1110

Stirpe F, Barbieri L (1986) Fine tuning of ribosomal accuracy. FEBS Lett 195:1–8

Uchida T, Egami F (1971) In: Boyer BD (ed) Enzymes, vol 4. Academic Press, London New York, pp 205–250

Veldman GM, Klootwijk J, Regt VCHF, Planta RJ, Branlant C, Krol A, Ebel J-P (1981) The primary and secondary structure of yeast 26S rRNA. Nucleic Acids Res 9:6935–6952

4 Effects on Gut Structure, Function and Metabolism of Dietary Lectins The Nutritional Toxicity of the Kidney Bean Lectin

Arpad Pusztai

4.1 Abstract

When fed to monogastric animals most of the dietary lectin, PHA, from the seeds of kidney bean *(Phaseolus vulgaris)* remains essentially unaltered and fully reactive during its passage through the entire digestive system. By binding to the surface cells of the gut PHA modifies their morphology, affects their normal function of digestion, absorption and secretion, increases wasteful protein and glycoprotein synthesis, speeds up cellular turnover, causes hypertrophy and hyperplasia in the small intestine and interferes with the proper workings of the local (gut) immune system. As a result of the combination of some or all of these effects the efficiency of the nutritional utilization of the diet and its protein content is seriously reduced.

Additionally the inclusion of pure PHA in the diet causes the involution of the thymus and an interference with the immune system. The size and the protein content of the liver is increased and the pancreas is also enlarged. Although it is not clear how this is related to the immediate drop in blood insulin concentration, the resulting increased lipolysis, muscle breakdown and mobilization of liver glycogen, combined with the disturbances of the normal functioning of the gut, have serious consequences for the growth and health of the animals. Thus through the effects of systemically absorbed PHA or indirectly through the reactions of PHA with gut endocrine cells, the body's hormone balance is shifted towards changes which favor increased rates of catabolic breakdown of all body components: lipid, protein and carbohydrate.

Animals fed on diets containing raw or inadequately cooked kidney beans suffer a rapid loss of weight and may eventually die, while after proper heat treatment the seeds are longer toxic (Johns and Finks 1920a, b). Soon after its discovery at the beginning of the century (Landsteiner and Raubitschek 1908) it was suggested that the hemagglutinin of these seeds was the factor responsible for the nutritional toxicity (Lüning and Bartels 1926).

Lectins are found in all foods and particularly in those of plant origin. However, the most extensively studied lectin is PHA from the seeds of the common bean *(Phaseolus vulgaris)* (Jaffé 1980; Pusztai et al. 1982, 1986; Pusztai 1985, 1986a, b). These seeds contain large amounts of lectin (about 10–15 % hemagglutinin of the total protein content) and this lectin has serious effects when included in the diets of monogastric animals. In this review an up-to-date and comprehensive assessment of our knowledge of the effects of dietary PHA on the digestive tract, its structure, functioning and metabolism, the resulting changes in systemic metabolism and the growth and health of the animals will be given. When these are sufficiently well known, references will also be made to the more fragmentary evidence available on the effects of other dietary lectins with particular emphasis on any similarities of differences in their mode of action.

The toxic factor in the kidney bean is now unequivocally established to be identical with the seed lectins and the extent of toxicity is shown to be dependent on the lectin content of the diets (Pusztai and Palmer 1977; Pusztai et al. 1981). All animals, including mature rats (Grant et al.

1985) and humans are affected (Noah et al. 1980; Bender and Reaidi 1982) and although the mechanism of the toxicity is by no means clear, its essential features are becoming more apparent. Thus in rats fed on diets containing PHA the utilization of dietary proteins is generally depressed. The resulting N deficiency of the body is further aggravated by the simultaneous increase in systemic catabolism. Above sufficiently high dietary lectin concentrations the animals are in negative N balance (Pusztai et al. 1981) and thus stop growing and eventually die. Accordingly, although the primary effects of the lectin are on the digestive tract and result in an interference with the digestion, absorption and utilization of dietary proteins, its secondary effects on systemic metabolism exacerbate the rapid loss of all body components and lead to a serious impairment of the growth and the health of the animal.

4.2   Effects on the Digestive Tract

All lectins examined so far show some degree of resistance to gut proteolysis. However, PHA is exceptionally resistant to proteolytic degradation (Pusztai et al. 1975, 1979; King et al. 1980a; Banwell et al. 1983; Hara et al. 1984) and as much as 90 % of the dietary lectin may survive passage through the entire gut (Pusztai 1980), while the other proteins of the diet are usually efficiently digested. Thus PHA reacts with the glycocalix and the surface membranes of a large proportion of the small intestinal cells and affects both the gross and the microscopic morphology of the gut (Pusztai and Greer 1984; Greer et al. 1985). Both mucosal and submucosal cells are affected (Pusztai 1987; Quarterman, Grant and Pusztai, unpublished). Protein, carbohydrate and DNA contents of these tissues increase and the extent of the increase is strictly dependent on the amounts of the dietary PHA. As a result of PHA binding to enterocytes and other cells (King et al. 1980a, 1986) brush border damage occurs in both rats (King et al. 1980a, b) 1982; Hara et al. 1983; Rossi et al. 1984) and pigs (King et al. 1983; Begbie and King 1985). An appreciable part of the bound lectin (in excess of 10 % of the dietary PHA) is then transported in endosomic vesicles and large endosomes into the gut cells and can eventually be traced to lysosomal compartments (King et al. 1986). Through its binding and intracellular transport PHA induces an accelerated cell loss, shortens and disorganizes small intestinal villi and reduces the absorptive surface of the gut (Pusztai 1986a; Pusztai et al. 1986).
Accordingly increased amounts of N are lost in the faeces and this includes not just undigested dietary N but also N of endogenous origin. Part of this is the result the mucotractive effect of PHA on goblet cells (Freed 1982) and the increased vascular permeability caused by the lectin affecting mast cells (Greer et al. 1985; Greer and Pusztai 1985). PHA also increases appreciably the protein synthesis rate of mucosal enterocytes (Palmer et al. 1987). Although the mechanism of this immediate "turn-on" by PHA of protein and glycoprotein synthesis is not entirely clear our preliminary results show that as it is preceded by an immediate and significant increase in tissue polyamine concentration (Wallace, Grant, Bardocz and Pusztai, unpublished), it is likely to be mediated through the well-known polyamine-generation pathway (Pusztai 1987). PHA also binds several of the brush border peptidases and thus may reduce the efficiency of amino acid and peptide transport through the absorptive gut epithelia (Rouanet and Besancon 1979; Rouanet et al. 1983; Triadou and Audran 1983; Erickson et al. 1985; Tajiri et al. 1986).
Increased amounts of undegraded material in the gut lumen facilitate bacterial overgrowth which, in turn, by competing for the same substrates, will further reduce the efficiency of absorption by the animal. PHA indeed causes a substantial proliferation on E. coli in the small intestine (Wilson et al. 1980; Banwell et al. 1985).

75

However, the PHA bound to enterocyte membranes may also directly anchor the *E. coli* to the gut through the bacteria's mannose-specific fimbrial lectins (Lis and Sharon 1986). The overall effects of dietary PHA on the gut are thus mediated either directly through its interference with the digestion and absorption of food or indirectly by stimulating a wastefully increased rate of protein synthesis, cellular turnover and proliferation, leading to a substantial reduction in the utilization of the diet with consequent poor growth and health.

4.3 Absorption of PHA from the Gut

A proportion of the dietary PHA endocytosed by brush border cells finds its way into the systemic circulation and internal organs of the body. Measurements of the amounts of ^{125}I-labelled PHA showed that within 3 h between 5–10 % of the toxic lectin initially introduced by stomach-tubing had become systemically absorbed (Greer 1983), while the amounts of the non-toxic tomato lectin absorbed under similar conditions were very small, i.e. less than 0.1 % of the initial dose (Kilpatrick et al. 1985).

Most of the absorbed PHA is initially bound to serum glycoproteins, however, with time an increasing amount becomes bound to various blood cells. At first, the binding is reversible but progressively more and more of it becomes irreversible as the PHA cannot be displaced by washing the cells with the haptenic glycoprotein, fetuin (Pusztai 1980).

Most recently it was found that systemic absorption of the dietary PHA is negligible in germ-free rats, although the lectin still binds to the brush border enterocytes and causes a disruption of its structure, similar to that found in conventional rats (Pusztai 1987). However, as in the absence of bacteria negligible amounts of PHA are taken up by these cells, the wastefully elevated cellular protein synthesis, hypertrophy and proliferation does not occur in them to the same extent as it does in their conventional counterparts (Pusztai 1987). Consequently, germ-free rats perform appreciably better nutritionally than rats which have a normal gut microflora (Jayne-Williams and Hewett 1972; Rattray et al. 1974; Wilson et al. 1980; Banwell et al. 1983).

One of the consequences of the systemic absorption of dietary PHA and its binding to various blood cells is that some of the lectin by binding to immunocompetent lymphocytes stimulates the production of humoral anti-lectin antibodies. All animals tested so far developed a powerful humoral antibody response of the IgG type exclusively to the dietary PHA (Pusztai 1980; Pusztai et al. 1981; Greer 1983; Williams et al. 1984; Begbie and King 1985; Grant et al. 1985). While the time course of the development shows the usual features of antibody production, generally its exclusiveness to the lectin argues strongly in favor of the great specificity of the lectin absorption from the gut and shows that the lectin has strong immunosuppressive effects against simultaneously introduced, unrelated antigens in the food. From the relatively extensive absorption of the dietary PHA into systemic circulation it is clear that perhaps because of the immunosuppressive effects of PHA on the gut immune system, the production of the local antibody of s-IgA may not be adequate. The bacterial proliferation in the small intestine, a secondary effect usually found in rats fed on diets containing PHA suggests that the s-IgA suppression may be a general phenomenon and not just directed towards the lectin.

Additionally, there is a mild anaphylactic effect on gut mast cells by the dietary PHA and this leads to the loss of serum proteins into the lumen of the gut (Pusztai and Greer 1984; Greer et al. 1985). Mice and rats also develop a systemic, immediatetype hypersensitivity to PHA mediated by the production of lectin-specific IgE (Pusztai et al. 1983b). All these immunological disturbances contribute to the generally poor nutritional performance and growth retardation of the animals fed on diets containing PHA.

4.4 Systemic Effects of the Absorbed PHA

Dietary PHA may bind to gut endocrine cells and interfere with gut hormone production thus indirectly affecting the systemic hormonal balance of the body. Alternatively, the systemically absorbed PHA may also directly interact with the internal endocrine organs of the body. The two processes may occur simultaneously. The increased extent of catabolic breakdown of body components observed may accordingly be the result of changes in systemic hormonal levels. Thus the increased rate of urinary urea output (Pusztai et al. 1981) and the extent of the generally elevated N catabolism was dependent on the amount of pure PHA in nutritionally good egg albumin-containing diets (de Oliveira 1986). When PHA is replaced by denatured lectin in these diets, the catabolic effects disappear. The dietary PHA also increased the losses of body lipid and glycogen (de Oliveira 1986; Pusztai 1986b, 1987; Pusztai et al. 1986). In fact, the rate of the disappearance of depot fat (mainly neutral triglycerides) exceeded that of the breakdown of body protein. Thus despite a net loss of body N the relative concentration of the body protein increased in line with the amount of dietary PHA. Furthermore, in the presence of 1% dietary PHA the concentration of 3-hydroxybutyrate in the urine increased five-fold over that of pair-fed controls (de Oliveira 1986). This showed that most of the effects on body lipid were to a large extent due to true lipid catabolism and not poor digestibility of the dietary lipid. A similar depletion of liver glycogen (but not of muscle) was also caused by the dietary PHA.

Indirect evidence for the involvement of hormonal mediation in the catabolic changes brought on by PHA was obtained when changes in the size of internal organs were investigated. One of the most conspicuous of such changes was the appreciable and pure PHA-dependent enlargement of the pancreas (de Oliveira 1986; Pusztai 1987) a tissue composed of cells originating from the same cells as the small intestine (also substantially enlarged due to the action of PHA). This was in fact the first time that such a change was shown to be due to a lectin. The trophic effect of PHA on the macroscopic structure of the pancreas may have far-reaching consequences for the hormone production of this organ. The previously observed severe reduction in circulating immunoreactive insulin concentration and the correspondingly increased blood glucagon levels may indeed be related to this PHA-induced change in the structure of the pancreas (Pusztai 1986; Pusztai et al. 1986). The reduction in blood insulin gives a plausible explanation for the muscle catabolism observed when rats were fed on diets containing PHA (Palmer et al. 1987; Pusztai 1987). Fractional rates of protein synthesis in muscles were reduced significantly in the hind leg muscles of rats fed on kidney bean proteins for 4 days, while protein degradation rates were unchanged with a resulting net loss of muscle mass.

Despite the abnormally low serum insulin concentrations in rats fed on diets containing PHA, these animals were not diabetic as their serum glucose concentration was almost normal and if anything it was slightly less than that found in the egg albumin control rats (de Oliveira 1986; Pusztai 1987). However, the abnormally energy-depleted state of the bean-fed rats was clearly shown after overnight fasting. As the liver glycogen stores of PHA-fed rats are nearly exhausted, the normal process of glycogenolysis on fasting does not generate enough blood glucose to replace that used up for metabolism. Additionally, because of the depletion of lipid reserves and reduction in muscle mass there is also a shortage of the substrates necessary for gluconeogenesis (glycerol and amino acids). Thus on fasting blood sugar concentration falls to perilously low values of about $0.7 \text{ mg} \cdot \text{ml}^{-1}$ in strong contrast to the relatively small reduction in blood glucose levels in the control rats fed on a diet containing no protein at all (de Oliveira 1986; Pusztai 1987). Thus the precise mechanism of all metabolic changes observed on feeding with PHA is not understood fully at present. While a number of the catabolic changes observed are consistent with the drastic reduction in insulin concentration caused by the dietary PHA, other changes such as those affecting the blood glucose concentration suggest the involvement of other hormones and a more complex regulation of the systemic metabolism of these animals.

4.5   Lectins from Other Food Plants

Although most foods of plant origin contain lectins in various amounts (Jaffé 1980; Nachbar and Oppenheim 1980; Gibbons and Danders 1981; Grant et al. 1983; Rea et al. 1985; Pusztai 1985, 1986a, b; Andersen and Ebbesen 1986; Liener 1986; Pusztai et al. 1986), only a few of them have been shown experimentally to be involved in antinutritive effects. However, what evidence exists appears to suggest the existence of a number of common features for the reaction mechanism of the toxic effects for most of the food lectins. Thus other members of the *Phaseolus* genus contain lectins very similar to those found in *Phaseolus vulgaris* (Grant et al. 1983; Pusztai et al. 1983a). These seeds or the proteins extracted from them when fed to rats were just as toxic as dietary PHA (Pusztai 1985, 1986a, b; Pusztai et al. 1986). The lectin from winged bean *(Psophocarpus tetragonolobus)* was also toxic and behaved very similarly to PHA. It showed a high resistance to gut proteolysis, was bound to intestinal epithelial cells causing severe disruption of the brush border membrane and impairing their proper digestive and absorptive function and interfered with the maturation process of these cells (Higuchi et al. 1983, 1984; Kimura et al. 1986).

Although there is no clear-cut evidence to implicate concanavalin A (Con A) in the nutritional toxicity of jack bean *(Canavalia ensiformis)* (Liener 1986), the binding of this lectin to the intestinal microvillous membrane *in vitro* has been shown to occur (Etzler and Branstrator 1974; Ichev and Ovtscharoff 1981). Moreover Con A introduced into intestinal loops caused lesions (Lorenzsonn and Olsen 1982; Ichev and Chouchkov 1983) very similar to those found with PHA. On the basis of this evidence, dietary lectins were suggested to be responsible for the normal turnover of brush border membranes (Lorenzsonn and Olsen 1982). The combined effects of the high resistance of Con A to gut proteolysis (Nakata and Kimura 1985) and the mucosal changes it caused (Lorenzsonn and Olsen 1982; Lorenz-Meyer et al. 1985) produced as serious interference with the normal functioning of the gut as previously found with PHA.

Rather curiously, although feeding soybeans is known to lead to inefficient utilization of the seed proteins, the evidence for, or against, the involvement of the seed lectin in poor performance is not conclusive (Liener 1986). However, soybean lectin is known to bind to intestinal microvillous membranes *in vitro* (Jindal et al. 1984). Moreover, a synergetic destabilizing effect of the lectin on rabbit jejunal epithelium with the consequent increased intestinal, permeability for toxic macromolecular antigens was shown to occur in the presence of seed saponins (Alvarez and Torres-Pinedo 1982; Torres-Pinedo 1983). These results may suggest that, at least in part, the lectin may make a contribution to the antinutritive effects of soybeans. The recent findings of the depression of circulating blood insulin concentration in rats fed on soybean proteins (Pusztai et al. 1986) may give further support to the involvement of the lectin in growth inhibition and other antinutritional effects.

Gluten lectins or wheat germ agglutinin (or both) may be one of the causative agents for the primary intolerance of the human gastrointestinal tract to peptides originating from wheat gluten in genetically predisposed human individuals (the so-called coeliac disease; Pusztai 1986a, b, 1987; Pusztai et al. 1986).

Castor bean lectin RCA_I is much less toxic than ricin, RCA_{II} when injected. However, on ingestion even ricin is far less toxic than by injection (Ishiguro et al. 1983). Indeed, its effect on the small intestine is mediated by the non-toxic, galactose-specific lectin B subunit which binds to the brush border, induces loss of villi, interferes with the absorptive epithelia and reduces sugar absorption from the gut (Ishiguro et al. 1983, 1984). All these effects are very similar to those found with PHA or Con A. The ricin-type toxic proteins display some modest selectivity in their reaction with mammalian cells with a definite preference for reaction with the transformed and cancerous cell types. The magic bullet concept of killing cancer cells with ricin-type toxins or their modified versions is being presently explored. However, this is outside the scope of the present review.

Finally, one more lectin merits mentioning, that is the lectin from the tomato fruit *(Lycopersicon esculentum)* which shows rather interesting properties. This lectin, like others discussed before, resists digestion in the gut and binds to intestinal villi. However, it does not destroy the structure of brush border membrane and only trace amounts of the lectin become absorbed by the epithelial cells (Kilpatrick et al. 1985). Consequently, the tomato lectin is not toxic. It is hoped that it may find some therapeutical use in coeliacs by blocking out intestinal sites and making them unavailable for the binding of the toxic gluten lectins.

In conclusion, some common features in the reaction mechanism of the nutritionally toxic food lectins are beginning to emerge:

- All toxic lectins are resistant to gut proteolysis.
- Toxic lectins react with and bind to small intestinal epithelial cell membranes, disrupt microvilli, increase cellular turnover and cell shedding and interfere with digestion and absorption.
- Toxic lectins affect the permeability of the gut.
- Toxic lectins may have an increased rate of systemic absorption.
- Toxic lectins may display systemic hormonal and other effects on the general metabolism and on the immune system.

Clearly, more work is needed, in particular, to investigate to what extent the findings of the systemic hormonal, immune and metabolic effects observed with the dietary PHA can be extended to the antinutritive effects observed with other dietary lectins and if these effects are truly representing common features in the nutritional toxicity of dietary lectins generally.

4.6   References

Alvarez JR, Torres-Pinedo R (1982) Interactions of soybean lectin, soysaponins, and glycinin with rabbit jejunal mucosa in vitro. Pediatr Res 16: 728–731

Andersen MM, Ebbesen K (1986) Screening for lectins in common foods by line-dive immunoelectrophoresis and by haemadsorption lectin test. In: Bøg-Hansen TC, Van Driessche E (eds) Lectins: biology, biochemistry, clinical biochemistry, vol 5. De Gruyter, Berlin (W), pp 95–108

Aub JC, Tieslau C, Lankester A (1963) Reactions of normal and tumor cell surfaces to enzymes. I. Wheat germ lipase and associated mucopolysaccharides. Proc. Natl Acad Sci USA 50: 613–619

Banwell JG, Boldt DH, Meyers J, Weber FL Jr (1983) Phytohemagglutinin derived from red kidney bean *(Phaseolus vulgaris)*: a cause for intestinal malabsorption associated with bacterial overgrowth in the rat. Gastroenterology 84: 506–515

Banwell JG, Howard R, Cooper D, Costerton JW (1985) Intestinal microbial flora after feeding phytohemagglutinin lectins *(Phaseolus vulgaris)* to rats. Appl Environ Microbiol 50: 68–80

Begbie R, King TP (1985) The interaction of dietary lectin with porcine small intestine and the production of lectinspecific antibodies. In: Bøg-Hansen TC, Breborowicz J (eds) Lectins: biology, biochemistry, clinical biochemistry, vol 4. De Gruyter, Berlin (W), pp 15–27

Bender AE, Reaidi GB (1982) Toxicity of kidney beans *(Phaseolus vulgaris)* with particular references to lectins. J Plant Foods 4: 15–22

Erickson RH, Kim J, Sleisenger MH, Kim YS (1985) Effect of lectins on the activity of brush border membrane-bound enzymes of rat small intestine. J Pediatr Gastroenterol Nutr 4:984–991

Etzler ME, Branstrator ML (1974) Differential localization of cell surface and secretory components in rat intestinal epithelium by use of lectins. J Cell Biol 62:329–343

Freed DLJ (1982) Lectins, allergens and mucus. In: Bøg-Hansen TC (ed) Lectins: biology, biochemistry, clinical biochemistry, vol 2. De Gruyter, Berlin (W), pp 33–43

Gibbons RJ, Danders J (1981) Lectin-like constituents of foods which react with components of serum, saliva and *Strep. mutans*. Appl Environ Microbiol 41:880–888

Grant G, More LJ, McKenzie NH, Stewart JC, Pusztai A (1983) A survey of the nutritional and haemagglutination properties of legume seeds generally available in the U.K. Br J Nutr 50:207–214

Grant G, Greer F, McKenzie NH, Pusztai A (1985) The nutritional response of mature rats to kidney bean *(Phaseolus vulgaris)* lectins. J Sci Food Agric 36:409–414

Greer F (1983) Local (intestinal) and systemic responses of animals to ingested *Phaseolus vulgaris* lectins: mechanism of lectin toxicity. Ph D thesis, Univ Aberdeen

Greer F, Pusztai A (1985) Toxicity of kidney bean *(Phaseolus vulgaris)* in rats: changes in intestinal permeability. Digestion 32:42–46

Greer F, Brewer AC, Pusztai A (1985) Effect of kidney bean *(Phaseolus vulgaris)* toxin on tissue weight and composition and some metabolic functions of rats. Br J Nutr. 54:95–103

Hara T, Tsukamoto I, Miyoshi M (1983) Oral toxicity of kintoki bean *(Phaseolus vulgaris)* lectin. J Nutr Sci Vitaminol 29:589–599

Hara T, Mukunoki Y, Tsukamoto I, Miyoshi M, Hasegawa K (1984) Susceptibility of kintoki bean lectin to digestive enzymes in vitro and its behaviour in the digestive organs of mouse in vivo. J Nutr Sci Vitaminol 30:381–394

Higuchi M, Suga M, Iwai K (1983) Participation of lectin in biological effects of raw winged bean seeds on rats. Agric Biol Chem 47:1879–1886

Higuchi M, Tsuchiya I, Iwai K (1984) Growth inhibition and small intestinal lesions in rats after feeding with isolated winged bean lectin. Agric Biol Chem 48:695–701

Ichev K, Chouchkov C (1983) Radioautographic study of leucine uptake and transmission of proteins by rat enterocytes following Concanavalin A binding. Acta Histochem 72:181–186

Ichev K, Ovtscharoff W (1981) Concanavalin-A binding sites on the intestinal microvillus membrane of rat. Acta Histochem 69:119–124

Ichiguro M, Mitarai M, Harada H, Sekine I, Nishimori I, Kikutani M (1983) Biochemical studies on the oral toxicity of ricin. I. Ricin administered orally can impair sugar absorption by rat small intestine. Chem Pharm Bull 31:3222–3227

Ishiguro M, Harada H, Ichiki O, Sekine I, Nishimori I, Kikutani M (1984) Effects of ricin, a protein toxin, on glucose absorption by rat small intestine (biochemical studies on oral toxicity of ricin II). Chem Pharm Bull 32:3141–3147

Jaffé WG (1980) Hemagglutinins (Lectins). In: Liener IE (ed) Toxic constituents of plant foodstuffs, chap 3, 2nd edn. Academic Press, London New York, pp 73–102

Jayne-Williams DJ, Hewitt D (1972) The relationship between intestinal microflora and the effects of diets containing raw navy beans *(Phaseolus vulgaris)* on the growth of Japanese quail *(Coturnix coturnix japonica)*. J Appl Bacteriol 35:331–334

Jindal S, Soni GL, Singh R (1984) Biochemical and histopathological studies in albino rats fed on soybean lectin. Nutr Rep Int 29:95–106

Johns CO, Finks AJ (1920a) The deficiency of cystine in proteins of the genus *Phaseolus*. Science 52:44

Johns CO, Finks AJ (1920b) Studies in nutrition. II. The role of cystine in nutrition as exemplified by nutrition experiments with the proteins of navy bean, *Phaseolus vulgaris*. J Biol Chem 41:379–389

Kilpatrick DC, Pusztai A, Grant G, Graham C, Ewen SWB (1985) Tomato lectin resists digestion in the mammalian alimentary canal and binds to intestinal villi without deleterious effects. FEBS Lett 185:299–305

Kimura T, Nakata S, Harada Y, Yoshida A (1986) Effect of ingested winged bean lectin on gastrointestinal function in the rat. J Nutr Sci Vitaminol 32:101–110

80

King TP, Pusztai A, Clarke EMW (1980a) Immunocytochemical localization of ingested kidney bean *(Phaseolus vulgaris)* lectins in rat gut. Histochem J 12:201–208

King TP, Pusztai A, Clarke EMW (1980b) Kidney bean *(Phaseolus vulgaris)* lectin-induced lesions in rat small intestine: 1. Light microscope studies. J Comp Pathol 90:585–595

King TP, Pusztai A, Clarke EMW (1982) Kidney bean *(Phaseolus vulgaris)* lectin-induced lesions in rat small intestine. 3. Ultrastructural studies. J Comp Pathol 92:357–373

King TP, Begbie R, Cadenhead A (1983) Nutritional toxicity of raw kidney beans in pigs. Immunocytochemical and cytopathological studies on the gut and the pancreas. J Sci Food Agric 34:1404–1412

King TP, Pusztai A, Grant G, Slater D (1986) Immunogold localization of ingested kidney bean *(Phaseolus vulgaris)* lectins in epithelial cells of the rat small intestine. Histochem J 18:413–420

Landsteiner K, Raubitschek H (1908) Beobachtungen über Hämolyse und Hämagglutination. Zentralbl Bakteriol Parasitenkd Infektionskr Hyg Abt 1 Orig 45:660–667

Liener IE (1986) Nutritional significance of lectins in the diet. In: Liener IE, Sharon N, Goldstein IJ (eds) The lectins, chap 10. Academic Press, London New York, pp 527–552

Lis H, Sharon N (1986) Lectins as molecules and as tools. Annu Rev Biochem 55:35–67

Lorenz-Meyer H, Roth H, Elsasser P, Hahn U (1985) Cytotoxicity of lectins on rat intestinal mucosa enhanced by neuraminidase. Eur J Clin Invest 15:227–234

Lorenzsonn V, Olsen WA (1982) In vivo responses of rat intestinal epithelium to intraluminal dietary lectins. Gastroenterology 82:838–848

Lüning O, Bartels W (1926) Über die Giftigkeit der Bohnen. Z Lebensm-Untersuch-Forsch 51:220–228

Nachbar MS, Oppenheim JD (1980) Lectins in the United States diet: a survey of lectins in commonly consumed foods and a review of the literature. Am J Clin Nutr 33:2338–2345

Nakata S, Kimura T (1985) Effect of ingested toxic bean lectins on the gastrointestinal tract in the rat. J Nutr 115:1621–1629

Noah ND, Bender AE, Reaidi GB, Gilbert RJ (1980) Food poisoning from raw red kidney beans. Br Med J 281:236–237

Oliveira JTA de (1986) Seed lectins: the effects of dietary *Phaseolus vulgaris* lectins on the general metabolism of monogastric animals. Ph D thesis, Univ Aberdeen

Palmer R, Pusztai A, Bain P, Grant G (1987) Changes in rates of tissue protein synthesis in rats induced in vivo by consumption of kidney bean lectins. Comp Biochem Physiol Ser B (submitted)

Pusztai A (1980) Nutritional toxicity of the kidney bean *(Phaseolus vulgaris)*. Rep Rowett Inst 36:110–118

Pusztai A (1985) Constraints on the nutritional utilization of plant proteins. Nutr Abstr Rev Ser B 55:363–369

Pusztai A (1986a) The biological effects of lectins in the diet of animals and man. In: Bøg-Hansen TC, Van Driesche E (eds) Lectins: biology, biochemistry, clinical biochemistry, vol 5. De Gruyter, Berlin (W), pp 317–327

Pusztai A (1986b) The role in food poisoning of toxins and allergens from higher plants. In: Robinson RK (ed) Developments in food microbiology, chap 7. Elsevier Appl Sci Publ, London New York, pp 179–194

Pusztai A (1987) Lectins. In: Cheeke PR (ed) Toxicants of plant origin, vol III. Proteins and amino acids. CRC Press, Boca Raton, Florida

Pusztai A, Greer F (1984) Effects of dietary legume proteins on the morphology and secretory responses of the rat small intestine. Protides Biol Fluids 32:347–350

Pusztai A, Palmer R (1977) Nutritional evaluation of kidney beans *(Phaseolus vulgaris)*: the toxic principle. J Sci Food Agric 28:620–623

Pusztai A, Grant G, Palmer R (1975) Nutritional evaluation of kidney beans *(Phaseolus vulgaris)*: the isolation and partial characterization of toxic constituents. J Sci Food Agric 20:149–156

Pusztai A, Clarke EMW, King TP (1979) The nutritional toxicity of *Phaseolus vulgaris* lectins. Proc. Nutr Soc 38:115–120

Pusztai A, Clarke EMW, Grant G, King TP (1981) The toxicity of *Phaseolus vulgaris* lectins. Nitrogen balance and immunochemical studies. J Sci Food Agric 32:1037–1046

Pusztai A, King TP, Clarke EMW (1982) Recent advances in the study of the nutritional toxicity of kidney bean *(Phaseolus vulgaris)* lectins in rats. Toxicon 20:195–197

Pusztai A, Croy RRD, Grant G, Stewart JC (1983a) Seed lectins: distribution, location and biological role. In: Daussant J, Mosse J, Vaughan J (eds) Seed proteins. Academic Press, London New York, pp 53–82

Pusztai A, Greer F, Silva Lima de Guia M, Prouvost-Danon A, King TP (1983b) Local and systemic responses to dietary lectins. In: Goldstein IJ, Etzler ME(eds) Chemical taxonomy, molecular biology and function of plant lectins. Liss, New York, pp 271–272

Pusztai A, Grant G, Oliveira JTA de (1986) Local (gut) and systemic responses to dietary lectins. IRCS Med Sci 14:205–208

Rattray EAS, Palmer R, Pusztai A (1974) Toxicity of kidney bean (Phaseolus vulgaris L.) to conventional and gnotobiotic rats. J Sci Food Agric 25:1035–1040

Rea RL, Thompson LU, Jenkins DJA (1985) Lectins in foods and their relation to starch digestibility. Nutr Res 5:919–929

Rouanet JM, Besancon P (1979) Effects d'un extrait de phytohemagglutinines sur la croissance, la digestibilite de l'azote et l'activite de l'invertase et de la (Na$^+$-K$^+$)-ATP-ase de la muqueuse intestinale, chez le rat. Ann Nutr Aliment 33:405–416

Rouanet JM, Besancon P, Lafont J (1983) Effect of lectins from leguminous seeds on rat duodenal enterokinase activity. Experimentia 39:1356–1358

Rossi MA, Mancini J, Filho, Lajolo FM (1984) Jejunal ultrastructural changes induced by kidney bean (Phaseolus vulgaris) lectins in rats. Br J Exp Pathol 65:117–123

Tajiri H, Lee PC, Lebenthal E (1986) Small intestinal mucosal hyperplasia caused by an enterokinase inhibitor from red kidney bean in rats. J Nutr 116:873–880

Torres-Pinedo R (1983) Lectins and the intestine. J Pediatr Gastroenterol Nutr 2:588–594

Triadou N, Audran E (1983) Interaction of the brush border hydrolases of the human small intestine with lectins. Digestion 27:1–7

Williams PEV, Pusztai A, Macdearmid A, Innes GM (1984) The use of kidney beans (Phaseolus vulgaris) as protein supplements in diets for young rapidly growing beef steers. Anim Feed Sci Technol 12:1-10

Wilson AB, King TP, Clarke EMW, Pusztai A (1980) Kidney bean (Phaseolus vulgaris) lectin-induced lesions in rat small intestine: 2. Microbiological studies. J Comp Pathol 90:597–602

Additions in Print

In recent morphological and immunofluorescence studies the involvement of PHA in toxic gut lesions in kidney bean fed rats has been further clarified (Bulajic et al. 1986). Thus, macroscopic examination of the gastro-intestinal tract of these animals indicated the occurrence of extensive inflammation. In addition to the previously extensively described binding of PHA to the mucosa of the proximal small intestine (King et al. 1980a, b, 1982, 1983, 1986; Hara et al. 1983; Rossi et al. 1984; Begbie and King 1985) pure PHA was now shown to bind also to the gastric epithelium (Bulajic et al. 1986). Moreover, in kidney bean-fed rats traces of PHA were found in the proximal large intestine bound to the mucosal coat. This extensive binding of the dietary lectin to the epithelium of the entire gastro-intestinal tract was suggested to lead to a number of pathological lesions in the gut (Bulajic et al. 1986), including the inhibition of brush border enzyme activities, increased shedding of epithelial cells, disturbances in the digestion and absorption of nutrients, bacterial overgrowth and interference with the immunoregulatory mechanism (Pusztai 1985, 1986a, b; Pusztai et al. 1986). Overall, for the rats fed on PHA-containing diets, these effects result in growth arrest, poor health and general toxicity.

One of the most striking effects of PHA is that, despite its overall toxicity, it acts as a growth factor for the small intestine (Pusztai and Greer 1984; Greer et al. 1985). This hyperplastic growth, previously observed in kidney bean-fed rats, has now been shown to be due wholly to the lectin content of these diets (Oliveira et al. 1988). Binding of PHA to membrane receptors and/or its subsequent endocytosis by epithelial cells (King et al. 1986) was shown to result in stimulation of cellular metabolism, protein and glycoprotein synthesis (Palmer et al. 1987) and increased contents of DNA and RNA (Pusztai and Greer 1984; Greer et al. 1985; Pusztai et al. 1986). This mucosal cell proliferation coincided with a substantial increase in the concentration of polyamines: putrescine, spermidine and spermine (Pusztai et al. 1988). Accordingly, PHA might utilize the same or similar polyamine-dependent pathways as used by other mucosal growth-stimulating signals in the small intestine (Hosomi et al. 1987; Luk and Yang 1987). Adaptive growth of the gut, which occurs in response to a number of stimuli, such as weaning, lactation, resection, obstruction or even simply just re-feeding after a period of starvation, requires the coincident de novo biosynthesis of putrescine *via* ornithine decarboxylase (ODC) (Hosomi et al. 1987; Luk and Yang 1987). In most instances the specific inhibition of the activity of this rate-limiting enzyme of polyamine metabolism with α-difluoromethylornithine (DFMO) leads to a simultaneous blockage of the growth process. The partial abrogation of the PHA-induced mucosal growth in the presence of DFMO (Pusztai et al. 1988) accordingly suggested a strong apparent dependence of the growth process on increased putrescine biosynthesis *via* ODC, similar to that occuring in adaptive growth. Thus the inclusion of DFMO (about 130–140 mg/rat/day) in PHA-containing diets was shown to reverse most of the increases in protein, DNA, RNA and polyamine contents found in rats fed on lectin-containing diets (Pusztai et al. 1988). Moreover, supplying exogenous polyamines through the diet or *via* blood circulation was shown to abrogate, in part, the inhibitory effects of DFMO on polyamine accumulation and the proliferative growth of the small intestinal tissue (Pusztai et al. 1988 and unpublished). Overall therefore it appears that the effects of PHA on gut epithelial cells *in vivo* are similar to its well-established *in vitro* mitogenic effects for peripheral lymphocytes (Kay and Cooke 1971). The maintenance of such a proliferative growth of the small intestine, however, requires appreciably increased amounts of amino acids and energy over and above the already high basal level requirements of the gut. In the presence of luminal PHA therefore a considerable part of the ingested food is diverted from the body and used for the metabolism of the gut epithelium. Accordingly, the nutritional toxicity of PHA, and possibly of other lectins, in addition to damaging the brush border, may also be due to their ability to interfere with membrane receptors involved in the signalling system of the small intestine and their subsequent endocytosis by epithelial cells. Against the background of general loss in body weight, by mimicking the effects of endogenous factors and activating the biochemical pathways of signal transduction through which the gut normally maintains its functionality and integrity, PHA induces and apparently wasteful growth of gut tissue at a high nutritional cost to the animals.

In addition to its previously observed interference with digestion in the small intestine by binding to and inhibiting several of the brush border peptidases (Rouanet and Besancon 1979; Rouanet et al. 1983; Triadou and Audran 1983; Erickson et al. 1985; Tajiri et al. 1986), PHA was shown to reduce the rate of in vitro digestibility of raw or heat-treated kidney bean proteins and several model proteins of animal origin (Thompson et al. 1986). Moreover, PHA also reduced the rate of digestibility of wheat or kidney bean starches *in vitro*, by non-competitively inhibiting both salivary and pancreatic amylases. Lectins, at concentrations normally encountered in foods, therefore, may be partly responsible for the observed health benefits derived from the slow digestion of foods, particularly in diabetics and obese people (Thompson and Gabon 1987).

More evidence for the direct involvement of PHA in the systemic toxic effects observed in rats on feeding with kidney bean-containing diets was provided by recent experimental work carried out with highly purified preparations of PHA (Oliveira et al. 1988). In addition to small intestinal hypertrophy and hyperplasia, pure PHA also induced pancreatic enlargement, increased liver weight, thymus atrophy and loss of skeletal muscle. The extent of all these changes was directly proportional to the daily intake of pure lectin in an otherwise nutritionally good diet. Moreover, a denatured PHA preparation caused no detectable changes in the same tissues (Oliveira et al. 1988). The overall effect of dietary PHA on systemic metabolism, as shown previously (Oliveira 1986; Pusztai 1986a, 1987; Pusztai et al. 1986; Palmer et al. 1987), is in the general direction of increased catabolism of tissue fat, protein and carbohydrate (Grant et al. 1987c). Most experimental evidence suggests that this wasting of body tissues is apparently the result of a modulation of the levels of a number of endocrine hormones, such as insulin, glucagon and, possibly glucocorticoids, caused by the presence of PHA in the gut lumen. As interference with the endocrine system is not confined to dietary PHA alone (Grant et al. 1987c), manipulation of hormone levels by dietary lectins may possibly open the way for potential pharmaceutical applications of orally given lectins in hormone therapy.

Lectins from other food plants

The involvement of the lectin (SBL) in the anti-nutritive effects observed on feeding raw soyabean to rats has now been firmly established (Grant et al. 1988). Thus, pure SBL included in nutritionally good diets inhibited the growth of rats and induced considerable changes in the physiology of the small intestine and the pancreas (Grant et al. 1987a, b, 1988). Similar to that found with PHA, SBL increased the size of the small intestine by inducing mucosal hyperplasia (Grant et al. 1987a, 1988). The proliferative growth was dependent on coincident increase in tissue polyamine concentration and was partially ameliorated by the inclusion of DFMO, an irreversible inhibitor of ODC (the rate-limiting enzyme of *de novo* polyamine biosynthesis). Feeding of diets containing pure SBL also caused pancreatic enlargement by both hypertrophy and hyperplasia (Grant et al. 1987b, 1988). The previously found depression of circulating insulin levels (Pusztai 1986; Pusztai et al. 1986; Grant et al. 1987c) in rats fed on soyabean proteins may, at least in part, be due to the effects of the lectin upon the structure of the pancreas and the consequent interference with the synthesis and/or secretion of the pancreatic islet hormones. Thus, the overall effects of the lectins from kidney bean and soyabean appear to show considerable similarities. It is, however, remarkable that while the effects of SBL on the small intestine and pancreas are quite similar to those of PHA and both lectins cause similar reduction in circulating insulin levels, dietary SBL does not induce the rapid weight loss observed when rats were fed on diets containing similar levels of the more toxic kidney bean lectin (Grant et al. 1988).

In contrast to the preferential binding of PHA to epithelial cells on the upper half of the villi of the proximal small intestine of rats (King et al. 1980a, b, 1982, 1983, 1986; Hara et al. 1983; Rossi et al. 1984; Begbie and King 1985; Bulajic et al. 1986; Pusztai 1986a; Pusztai et al. 1986) concanavalin A (Con A; with specificity for D-mannose/D-glucose) and wheat germ agglutinin (WGA; with specificity for N-acetyl-D-glucosamine oligomers) affected enterocytes at the base of the villi rather than those at the top (Sjolander et al. 1986). This is in accord with the sugar specificity of these lectins and the known preponderance of high mannose-containing carbohydrate structures in membrane glycoconjugates of the less differentiated epithelial cells in the proliferative compartment, the crypts of the small intestine. Lectin-binding led to extensive disarrangement of the cytoskeleton of the crypt cells, shortening of the microvilli and increased endocytosis of the bound lectins by these cells (Sjolander et al. 1986). These morphological changes leading to increased uptake of macromolecules through the intestinal wall into systemic circulation bear more than a superficial resemblance to those found in coeliac patients, thus supporting the involvement of lectins or lectin-like substances in the pathogenesis of this disease (Sjolander et al. 1986). Similar effects of concanavalin A were observed in the gastrointestinal tract of neonatal guinea pigs (Weaver and Bailey 1987). In addition to the severe mucosal damage caused to the stomach by the intrapharyngeally infused concanavalin A, the

84

lectin reduced the mean villus height and increased the mean crypt depth in the jejunum. Crypt cell production rates were increased by almost fivefold and the permeability of the gut to the lectin and other large molecules and particles was substantially increased. These findings have again implicated the involvement of lectin-like substances in the aetiology of childhood coeliac disease (Weaver and Bailey 1987).

Finally, based largely on previous studies with PHA, the main features of the reaction of toxic food lectins with the gastrointestinal tract have already been established. It now appears that this reaction mechanism may be generally applicable to most lectins from either higher or lower plants. Thus, in addition to the usually homotetrameric lectins, which are made up of lectin subunits only (PHA, SBL, concanavalin A), most A (toxic subunit) – B (lectin subunit) type plant or microbial toxins are also endocytosed by gut epithelial cells extensively (Aizpurua and Russel-Jones 1988; Pusztai 1988). However, even in these A-B toxins, which are absorbed via subunit dissociation (Waksman et al. 1980), endocytosis is effected through the lectin part of the molecule. In this process the lectinic B subunit is first attached to membrane receptors and its hydrophobic part is inserted as a channel (Blanstein et al. 1987) into the membrane through which the toxic A subunit, still attached by disulphide bridges to the B-chain, enters the cell. Reduction of the cystine residue holding the two subunits together occurs in the Golgi stacks after which the free A-subunit is liberated (Deurs et al. 1988). Accordingly, the extraordinary biological effectiveness of all types of lectins is the result of their specific interaction with glycoconjugates, whose presence in eukaryotic cell membranes is obligatory, and their subsequent receptor-mediated endocytosis by these cells. Stabilized against breakdown by their attachment to membranes, the endocytosed lectins survive the passage through the epithelial cells in appreciable quantities. A part of the intracellularly absorbed lectin may then be released into the cytoplasm and/or expelled from these cells through the basolateral membranes into systemic circulation. In either case, through this highly specific cellular uptake, lectins may gain access to tissues and systemic circulation which are normally protected against such direct exposure to other food proteins. As, at the same time, most lectins have diverse and strong biological activities, their profound effects on metabolism, growth and the health of man and animals are beginning to be widely appreciated.

References

Citations which were referred previously are not included in this reference list.

Aizpurua H J de, Russel-Jones G J (1988) Oral vaccination. Identification of classes of proteins that provoke an immune response upon oral feeding. J Exp Med 167:440–451

Blanstein R O, Germann W J, Finkelstein A, DasGupta B R (1987) The N-terminal half of the heavy chain of botulinum type A neurotoxin forms channels in planar-phospholipid bilayers. FEBS Lett 226:115–120

Bulajic M, Cuperlovic M, Movsesijan M, Borojevic D (1986) Interaction of dietary lectin (phytohaemagglutinin) with the mucosa of rat digestive tract – Immunofluorescence studies. Periodicum Biologorum 88:331–336

Deurs B van, Sandvig K, Petersen O W, Olsnes S, Simons K Griffiths G (1988). Estimation of the amount of internalized ricin that reaches the trans-Golgi network. J Cell Biol 106:253–267

Grant G, Watt W B, Stewart J C Pusztai A (1987a) Changes in the small intestine and hind leg muscles of rats induced by dietary soyabean *(Glycine max.)* proteins. Med Sci Res 15:1355–1356

Grant G, Watt W B, Stewart J C Pusztai A (1987b) Effects of dietary soyabean *(Glycine max.)* lectin and trypsin inhibitors upon the pancreas of rats. Med Sci Res 15:1197–1198

Grant G, de Oliveira J T A, Dorward P M, Annand M G, Waldron M Pusztai A (1987c) Metabolic and hormonal changes in rats resulting from consumption of kidney bean *(Phaseolus vulgaris)* or soyabean *(Glycine max.)*. Nutr Rep Int 36:763–772

Grant G, Watt W B, Stewart J C Pusztai A (1988) Local (intestinal) and systemic responses to dietary soyabean lectin. In: Freed DLJ, Bøg-Hansen TC (eds) Lectins: biology, biochemistry, clinical biochemistry, vol 6. Sigma Library

Hosomi M, Lirussi F, Stace N H, Vaja S, Murphy G M Dowling R H (1987) Mucosal polyamine profile in normal and adapting (hypo- and hyperplastic) intestine: effects of DFMO treatment: Gut 28:S1, 103–107

Kay J E Cooke A (1971) Ornithine decarboxylase and ribosomal RNA synthesis during the stimulation of lymphocytes by phytohaemagglutinin. FEBS Lett 16:9–12

Luk G D Yang P (1987) Polyamines in intestinal and pancreatic adaptation. Gut 28:S1, 95–101

Oliveira J T A de, Pusztai A Grant G (1988) Changes in organs and tissues induced by feeding of purified kidney bean *(Phaseolus vulgaris)* lectins. Nutr Res (in press)

Pusztai A (1988) Transport of proteins through the membranes of the adult gastrointestinal tract – A potential for drug delivery. Adv Drug Delivery Rev (in press)

Pusztai A, de Oliveira J T A, Bardocz S, Grant G Wallace H M (1988) Dietary kidney bean lectin-induced hyperplasia and increased polyamine content of the small intestine. In: Freed DLJ, Bøg-Hansen TC (eds) Lectins: biology, biochemistry, clinical biochemistry, vol 6. Sigma Library

Sjolander A, Magnusson K E Latkovic S (1986) Morphological changes of rat small intestine after shorttime exposure to concanavalin A or wheat germ agglutinin. Cell Struct Funct 11:285–293

Thompson L U Gabon J E (1987) Effect of lectins on salivary and pancreatic amylase activities and the rate of starch digestion. J Food Sci 52:1050–1053

Thompson L U, Tenebaum A V Hui H (1986) Effect of lectins and the mixing of proteins on rate of protein digestibility. J Food Sci 51:150–152

Waksman A, Hubert P, Cremel G, Rendon A Burgun C (1980) Translocation of proteins through biological membranes. A critical view. Biochim Biophys Acta 604:249–296

Weaver L T Bailey D S (1987) Effect of the lectin concanavalin A on the neonatal guinea pig gastrointestinal mucosa in vivo. J Pediatr Gastroenterol Nutr 6:445–453

5 Potential Participation of Tumor Lectins in Cancer Diagnosis, Therapy and Biology

Hans-Joachim Gabius

5.1 Introduction

Cancer is among the leading causes of death. Although undisputed progress has been made in the last decades in the war against cancer, our knowledge of the clinical and biological behavior and property of many types of cancer remains, for the time being, largely empirical (Cairns 1985; Cohen and Diamond 1986). It is thus a mandatory obligation of basic and applied research to continuously refine and expand the possibilities for accurate histological assessment, for proper therapeutic management and for elucidation of processes in tumor biology as a basis to develop rational ways, e. g., of interfering with tumor growth and spread and controlling tumor differentiation. In this respect, shedding light on recognitive systems, by which biological information that is pivotal to the regulation of cellular communication, growth, and differentiation under normal or pathological conditions is stored and decoded, appears to be an attractive choice on the way to pinpointing relevant molecular characteristics. The highly specific fit of a ligand with a receptor provides a modular regulatory element that the evolutionary process has repeatedly and reliably used in different types of cells for many different purposes. Whereas enzyme-substrate or antibody-antigen interactions cover the interactions of diverse molecular types of ligands with proteins as receptors, regulatory peptides or peptide domains within larger proteins exert their biosignaling functions by selectively binding to protein receptors. In these cases ligand and receptor function is made up in the language of amino acids.

When considering the mediation of recognition phenomena there is still a strong prejudice towards thinking only in terms of proteins. Beyond this means of amino acid sequences, chain structures of another common part of cellular biochemistry fulfill the prerequisites to be able to confer biological information. Monosaccharides may serve as letters in an additional vocabulary of biological information. Besides variability in the primary sequence of carbohydrate chains, their individual linkage type, and the possibility of incorporating branch points add to the size of the monosaccharide-based vocabulary. Such chains of monosaccharides are well-known constituents of glycoconjugates, although their precise functions have so far remained elusive (Hakomori 1981; Berger et al. 1982; Sairam 1985; Cook 1986). Conformational analysis of the carbohydrate part of glycoconjugates has further corroborated the assumption that they can act as ligands in biosignaling. Hereby, it has become evident that a certain structural organization of oligosaccharide chains allows them to adopt distinct conformations, forming unique and accessible spatial structures on the surface of glycoconjugates, as required in biosignaling (Brisson and Carver 1983; Montreuil 1986; Homans et al. 1987).

Having defined the ligand part of a prospective recognition system, respective receptors should be present to complete it. In molecular terms, these receptors can be specific carbohydrate-binding proteins such as endogenous lectins. Protein-carbohydrate interactions, established by lectin-carbohydrate recognition, may thus be a crucial manner of participating in the complex processes of fertilization, tissue development, growth, and differentiation (Barondes 1986; Lis and Sharon 1986; Gabius 1987a). Regulation of the interplay of endogenous lectins and their carbohydrate ligands, with all required variability afforded to this interaction by the heterogeneity of both classes of molecules, places a further level of modulatory possibilities at any cell's disposal. Due to the presence and differential expression of different types of cellular glycoconjugates in normal development and malignant transformation, monitored by plant lectins or monoclonal antibodies as sugar-specific probes (Muramatsu 1984; Feizi 1985; Caselitz 1987; Damjanov 1987), their informational content resides not only in their individual expression, but also in the pattern of the different present structures and in the presence of corresponding receptors. Compared to the knowledge about the regulation of the carbohydrate structures of cellular glycoconjugates, much less is known about the receptor part of this code system. Individual ligand-receptor pairs, however, will definitely form the characters of the code, governed by protein-carbohydrate interactions. They cause respective biological effects in the contextual interplay with other biosignals, received by cells.

This lack of information about endogenous lectins prompts the investigation of their presence and regulation, especially in tumors, in the hope of being able to apply this knowledge to advantage in clinical practice. Tumor lectinology, inevitably interdisciplinary, encompasses the fields of preparative carbohydrate chemistry, biochemistry, cell biology, immunology, basic and clinical oncology and pathology (Gabius et al. 1987a). This brief review is not intended to be comprehensive, but to thoroughly illustrate main lines of development for lectin-based refinements in diagnostic and therapeutic procedures.

5.2   Is Histopathological Detection of Endogenous Sugar Receptors Feasible and Diagnostically Meaningful?

Since the structures of the carbohydrate chains of cellular glycoconjugates contribute toward establishing main determinants of cellular identity (Feizi and Childs 1987), plant lectins have proven their inconfutable merit as investigative indicators for alterations in tumor pathology (Coggi et al. 1983; Cooper 1984; Allison 1986; Caselitz 1987). Indeed, the fucose-specific lectin UEA-I has received general acceptance as a reliable marker for tumors of vascular origin

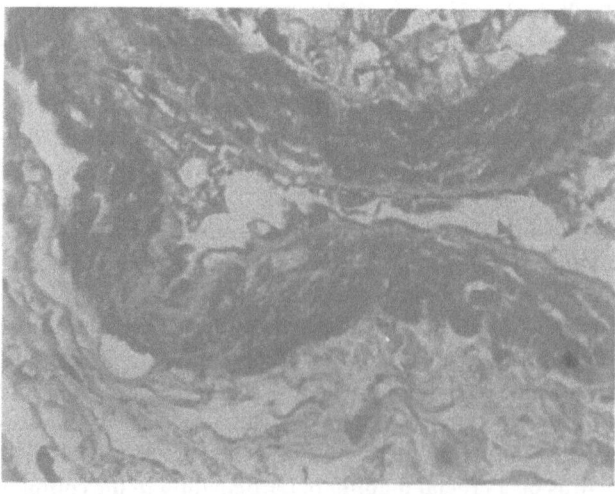

Fig. 5.1 Staining of smooth muscle cells of a venous blood vessel in normal breast tissue by biotinylated, fucosylated bovine serum albumin and ABC reagents; magnification 428 x

(Miettinen et al. 1983). However, any exogenously added probe will gain insight into only one aspect of the above type of cellular informational code system, based on protein-carbohydrate interactions. To localize the endogenous receptor part to carbohydrate ligands, it is necessary to develop appropriate tools and to standardize their practical handling. Whereas the number of immunohistochemical studies with antibodies to endogenous lectins in tumors is at present rather restricted (Gabius et al. 1986a), evaluation of another class of tools to assess the pattern of endogenous sugar receptor expression histopathologically within routine procedures is more advanced. Synthetic conjugates of a suitably labeled, itself histochemically inert carrier protein with covalently attached, histochemically crucial carbohydrate moieties in defined and desired amounts, operationally termed neoglycoproteins (Stowell and Lee 1980; Aplin and Wriston 1981), offer stimulating potential in this respect. Their easy and multifarious synthetic access allows the establishment of a panel of conjugates. Their application is exemplarily demonstrated for visualization of fucose-specific receptors in a human venous blood vessel by a

Table 5.1 Summary of (neo)glycoprotein staining of normal, non-malignant and malignant breast tissues

Type of lesion	Labelled (neo)glycoprotein								
	lac-BSA (5)[3]	lac BSA (17)[3]	ASF	ATF	ALF	man-BSA (4–6)	glcNAc-BSA (4–6)	fuc-BSA (4–6)	galNAc-BSA (4–6)
Normal breast (2)	–	N	N	N+	N	–	iN+	N	–
	–	–	C	C	C	–	–	(C)	–
Fibrocystic desease (7)	hv	iN	N	N	N	N	iN	N	–
	hv	–	(C)	(C)	C	–	–	(C)	–
Fibroadenoma (1)	–	–	N+	–	(N)	N+	N	N	–
	–	–	(C)	(C)	C	(C)	–	C	–
Invasive ductal ca[1] (22)	hv	N+	(N)	(N)	N-N+	hv	N+	V	hv
	hv	(C)	C	C-C+	C-C+	hv	(C)	V	hv
Scirrnous ductal ca (1)	N	–	(N)	(N)	–	N	N+	–	–
	(C)	–	C	C	C	–	(C)	–	–
Medullary ca (1)	–	–	N	(N)	N	N	(N)	(N)	–
	–	–	C	C	–	–	–	–	–
Lobular ca (3)	–	–	(N)	(N)	N	N	(N)	(N)	–
	–	–	C	C+	C+	C+	(C)	(C)	–
CLIS[2] (1)	–	(N)	–	(N)	N	(N)	–	N	–
	–	(C)	–	(C)	C	C	–	C+	–
Papillary ca (1)	–	N	(N)	(N)	N	N	N	–	–
–	(C)	C+	C+	C+	C	–	–	–
Comedo ca (1)	–	(N)	–	–	iN	iN+	(N)	(N)	–
	–	(C)	C	(C)	C+	C	–	(C)	–
Mucinous ca (2)	–	–	N	(N)	N	N	(N)	(N)	–
	–	C	C+	C+	C+	C	(C)	C+	–

X+, X, (X): degrees of staining, N: nucleus and nuclear membrane, C: cytoplasm and, graded, cellular membrane, hv, v: highly variable and variable staining intensity within a specimen, iN: staining of only individual nuclein within a specimen; the first row within each series represents nucleus staining, the second row represents cytoplasmic staining,

[1]: carcinoma, [2]: carcinoma lobulare in situ, [c]: sugar residues per carrier molecule: sugars: lactose (lac), mannose (man), N-acetylglucosamine (glcNAc), fucose (fuc) and N-acetylgalactosamine (galNAc); glycoproteins: asialofetuin (ASF), asialotransferrin (ATF) and asialolactoferrin (ALF); BSA: bovine serum albumin

89

fucosylated and biotinylated carrier protein (Fig. 5.1). Furthermore, naturally occurring glycoproteins, deblocked by removal of sialic acid residues from the termini of the carbohydrate chains, can be chosen to expand the armamentarium to complex carbohydrate sequences as ligands, for complex structures are only gradually becoming accessible by chemical synthesis. Again, biotinylation renders these probes detectable in tissue sections by standard protocols. Controls to ascertain the carbohydrate-specific binding and to exclude any contribution of non-specific protein-protein interaction, exerted by the carrier protein, validate the potential value of these tools in histopathology in various studies of different tumor systems (Bardosi et al. 1988a, b; Gabius et al. 1988a; Kayser and Gabius 1988).

Panels of such tools to localize endogenous sugar receptors in tumor tissue sections prove their potential value as discriminating diagnostic reagents, as exemplified for breast tumors (Table 5.1) and various types of brain tumors (Table 5.2). Initial assessment of their binding pattern, referring to lung cancer, reveals pronounced differences between small cell lung cancer and other major types of non-small cell lung cancer, as well as between carcinomatous cells and inflammatory cells and pneumocytes (Table 5.3a, b). When results of different individual mar-

Table 5.2 Cytoplasmic (Cp) and nuclear (N) carbohydrate-binding capacity (Σn)

Receptor specificity	Type of (neo)glyco-protein	Oligodendroglioma differentiated (Σn)		anaplastic (Σn)		Ganglio-cytoma (Σn)		Subtypes of meningiomas (Σn)				
		Cp	N	Cp	N	Cp	N	A	B	C	D	E
	lac (diazo)-BSA	6	+	5	+	6	+	4	6	4	3	3
	lac (red. am.)-BSA	1		2		4		2	0	2	0	0
β-Galactoside	gal-β-(1,3)-glcNAc-BSA	3	+	1		0		0	0	0	0	0
	asialotransferrin	1		2		5		3	3	1	3	1
	asialofetuin	6		6		6	+	4	4	6	6	3
	asialolactoferrin	1		0		0		1	3	1	4	3
α-Galactoside	melibiose-BSA	0		0		0		1	0	0	0	0
N-Acetylated sugars	N-acetyl-glucosam.-BSA	5	+	3		4		5	3	3	2	3
	N-acetyl-galactosam.-BSA	0		0		0		0	0	0	2	4
α-Glucoside	maltose-BSA	3		0		4		4	3	4	2	0
β-Glucoside	cellobiose-BSA	3		3		0		0	0	0	0	0
α-Fucoside	fucose-BSA	4	+	6	+	5		2	2	5	3	3
α-Mannoside	mannose-BSA	5		4		6		6	4	6	3	3
Phosphorylated sugars	mannose-6-P-BSA	4	+	4	+	5		3	6	6	3	3
	galactose-6-P-BSA	5	+	2		3		3	2	1	0	1
Sugars with a	sialic acid-BSA	0		0		0		0	0	0	0	0
carboxyl group	glucuronic acid-BSA	1		1		3	+	4	0	0	0	1
β-Xyloside	xylose-BSA	6	+	6	+	6		6	6	6	5	2
General binding indexes	$\Sigma\Sigma n = $	54		45		57		48	42	45	36	30

Staining intensity: Σn = 5–6: strong, 4–3: medium, 2–1: weak, 0: no staining; subtypes of meningiomas: meningotheliomatous, B: fibroblastic, C: angioblastic, D: submalignant, E: malignant, lac: lactose, diazo: synthesized by diazotation, red. am.: reductive amination, gal: galactose, glcNAc: N-acetylglucosamine

Specificity	Tumor type			Table 5.3a Neoglycoprotein binding to tumor cells in 10 cases of each lung cancer type
	Small cell	Adenocarcinoma	Epidermoid	
Lactose	0[1]	4	4	
Melibiose	0	2	2	
Mannose	0	7	5	
Fucose	(1)	8	6	
Maltose	0	9	5	
Fucoidan	0	0	0	

[1]: Number of positive cases

Table 5.3b Neoglycoprotein binding to accompanying inflammatory cells and pneumocytes in 30 cases of lung cancer

Specificity	Cell type				
	Macro phages	Granulo cytes	Plasma cells	Lympho cytes	Pneumo cytes
Lactose	0[1]	0	0	0	0
Melibiose	12	0	4	0	0
Mannose	12	0	0	3	0
Fucose	0	0	8	0	0
Maltose	9	3	6	3	7
Fucoidan	0	18	7	4	0

[1]: Number of positive cases

kers are combined, the accuracy of neoglycoprotein-based diagnosis can be elevated, although a notable degree of heterogeneity within tumor specimens can also be detected with this marker type (Kayser and Gabius 1988). Since further initial encouraging observations have been obtained with this methodology in dermato-histopathology (Berger, pers. commun.), refinement of these tools to meet the chief requirement of surgery-associated histopathological evaluation for performance in a minimum number of steps is compulsory. Preparation of integral neoglycoprotein-enzyme conjugates meets this practical demand, employing heterobifunctional reagents like N-succinimidyl-3-(2-pyridyldithio) propionate, that couple the two types of molecules via a disulfide-bridge (Gabius et al. 1987b). Gel electrophoretic analysis in the absence and in the presence of a reductive agent, breaking this connecting disulfide link, assures adequate selection of the desired monoconjugate after separation of the various reaction products, as shown in Figure 5.2A, B.

Together with plant lectins, disclosing localization of defined carbohydrate parts of glycoconjugates, neoglycoproteins permit inferring histochemical information on the recognitive system of endogenous sugar receptors and their carbohydrate ligands. Although technical advances will have to address question concerning standardization and optimalization in the preparation and utilization of the probes and the processing of tissue sections, their application will definitely not be restricted to histopathology. Hepatological, hematological, and neurochemical studies bear witness to more widespread applicability (Kolb-Bachofen et al. 1982; Kataoka and Tavassoli 1985; Gabius et al. 1988b), as illustrated for various neoglycoproteins and their binding to different parts of human sural nerves in Table 5.4.

91

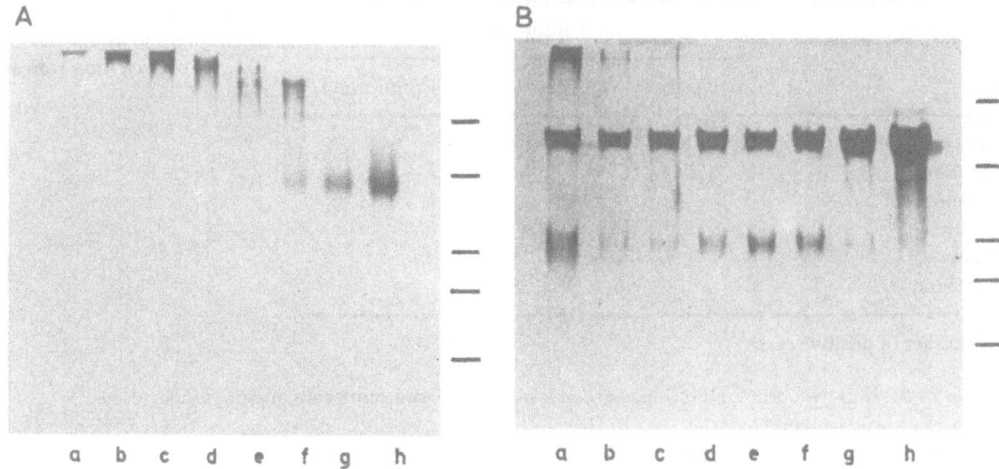

Fig. 5.2 SDS-polyacrylamide gel electrophoresis in the absence (A) and presence (B) of reductive agent of combined fractions. The fractions resulted from separation of the reaction mixture with thiol-containing peroxidase and 2-pyridyl disulfide-containing bovine serum albumin, given in the order of elution. Standards for molecular weight designation, indicated by bars, are phosphorylase b (MW 97,000), bovine serum albumin (MW 66,000), egg albumin (MW 44,000), glyceraldehyde-3-phosphate dehydrogenase (MW 36,000) and carbonic anhydrase (MW 29,000). Standard proteins have been generally treated with reductive agent

As the carbohydrate part of cellular glycoconjugates is increasingly being considered not as a passive, inert component, but as an information-bearing modulatory element, sulfated polysaccharides, prominent constituents of proteoglycans, are correspondingly adduced to profoundly participate in various physiologically eminent processes (Höök et al. 1984; Iozzo 1985; Hascall 1986). Histochemical localization of respective endogenous receptors is facilitated by biotinylation of sulfated polysaccharide structures to produce the necessary probes. Their usefulness is exemplarily shown for labeled heparin in Figure 5.3. This technique confirms that receptors specific for sulfated polysaccharides appear to be differentially expressed among various cell types and developmentally regulated (Debbage et al. 1988). The placenta, as an organ of embryonic origin with a limited number of cell types and undergoing salient changes in the course of development, solidly emphasizes such developmental changes for this receptor class (Table 5.5). Tumor-relevant studies will be of particular interest in this context, because growth factors and angiogenetic factors belong to this group of receptors.

In aggregate, the current status of application of the described tools to localize endogenous sugar receptors in tumor tissue sections encourages to suggest that this field is likely to gain added momentum in the near future. Besides enabling a step towards elucidating the functional significance of endogenous sugar receptors, such results will also serve as a valuable guideline for deliberate selection of certain carbohydrate specificities of endogenous sugar receptors for biochemical analysis, e. g., in cases of differences in normal progenitor cell-tumor cell comparison. The principally proven feasibility to employ labeled neoglycoproteins and sulfated polysaccharides in histopathology additionally justifies daring to suggest their advantageous exploitation in tumor therapy.

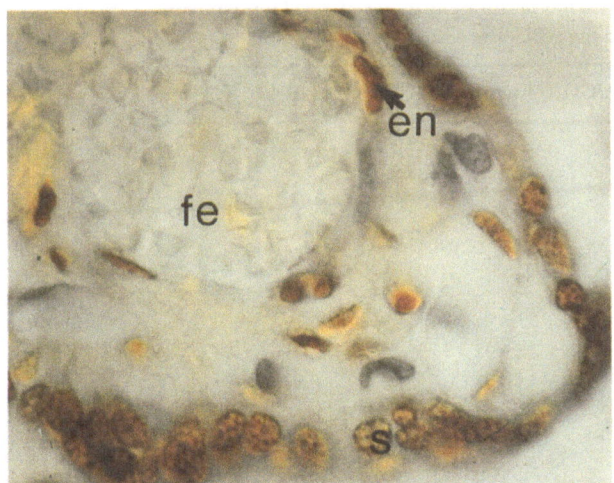

Fig. 5.3 Staining of a section of human placenta at term by biotinylated heparin. The syncytiotrophoblast nuclei are intensely labeled (s). Fetal erythrocytes (fe) are unlabeled, but an endothelial nucleus (en) is labeled; magnification 1000 ×

Table 5.4 Staining of human peripheral nerves with biotinylated (neo)glycoproteins

Type of (neo)glycoprotein	Part of nerve			
	Endoneurium	Epineurium	Myelin	Axoplasm
lac(diazo)-BSA	−	−	++	++
lac(red. amin)-BSA	−	++	+−++	++
gal-β1,3-glcNAc-BSA	−	+	+	−
ASF	+	++	++	+
ATF	−	−	+	−
ALF	+	+	+	+
mel-BSA	−	−	−	−
galNAc-BSA	−	+	++	−
glcNAc-BSA	++	++	++	++
mal-BSA	−	+	++	−
cel-BSA	−	+	+	+
fuc-BSA	++	++	+−++	++
man-BSA	−	+	++	+
man-6-P-BSA	+	++	++	++
gal-6-P-BSA	−	+	−	−
Sialic acid-BSA	−	−	−	−
Glucuronic acid-BSA	−	−	−	−
xyl-BSA	−	+	+	++
Fucoidan	−	−	−	−
Fucoidan (desulfated)	−	−	+	−

lac: lactose, diazo: synthesized by diazotation, BSA: bovine serum albumin, red. amin: synthesized by reductive amination, gal: galactose, glcNAc: N-acetylglucosamine, ASF: asialofetuin, ATF: asialotransferrin, ALF: asialolactoferrin, mel: melibiose, galNAc: N-acetylgalactosamine, mal: maltose, cel: cellobiose, fuc: fucose, man: mannose, P: phosphate, gal: galactose, xyl: xylose

Table 5.5 Staining of human placental tissues at different stages of development with biotinylated sulfated polysaccharides and the ABC procedure

		HEP	FUCO	Fuco (s)	DERM	CAR-k	CAR-1
Dedidua (8 weeks)		−/+	−	−	−	−	−
Syncytiotrophoblast (8 weeks)	N	−	−	−/+	−	−	−
	C	−/+	−	−	−	+/++	+/++
Cytotrophoblast (8 weeks)	N	−/++	−	−/+	−	++	−
	C	−/++	−	−	−	+/++	+/++
Endothelium (8 weeks)		−	−	−	−	+/++	−
Macrophages (8 weeks)	M	−/++++	−	(++)	−	−	−
	F	−/++++	−	(++)	(+)	(++)	−
Leucocytes (8 weeks)	M	(++)	++	−/++	−/++	−	−/+
	F	(++)	++	−/++	−/+	−	−
Erythrocytes (8 weeks)		−	−	−	−	+/++	NU + AN −
Decidua (Term)	N	+++	+/++	−	−	−	−
	C	+/++					
Syncytiotrophoblast (Term)	N	−/+++	−/+	−	−	−	−
	C	−/+	−	−	−	−	−
Macrophages (Term)	M	++++	(+++)	(+++)	(+++)	(++)	(+++)
	F	++++	(+++)	(+++)	(+++)	(++)	(+++)
Endothelium (Term)	N	+++	−/+	−	−	−	−
	C	−	−	−	−	−	−
Leucocythes (Term)	M	(++)	++	++	++	++	−/++
	F	(++)	++	++	++	−/+++	−/++
Erythroblast (Term)		−/+	−	−	−	−	−

HEP: heparin, FUCO: fucoidan, Fuco(s): desulfated fucoidan, DERM: dermatan sulfate, CAR-k: kappa-carrageenan, CAR-l: lambda-carrageenan
C: cytoplasm, F: fetal, M: maternal, N: nucleus, Nu/An: nucleated/anucleated fetal erythrocyte

5.3 Can Neoglycoproteins Serve as Vehicles for Therapeutic Agents in Targeted Drug Delivery to Tumor Cell Lectins?

Cancer chemotherapy provides variably effective treatment for the various forms of cancer. The clinical use of chemotherapeutic agents, however, can only be vindicated if its beneficial effects outweigh its toxic side effects, because inhibition of tumor growth is inevitably accompanied by serious toxicity to rapidly proliferating normal cells. Restricting the access of chemotherapeuticals to non-target cell types and augmenting their amount and persistence in the vicinity of cancer cells comprise the aims of controlled and targeted drug delivery (Poznansky and Cleland 1980; Poste 1983; Freeman and Mayhew 1986).

Among the potential carriers for tumoricidal agents, plant lectins historically began to be exploited even before the advent of monoclonal antibodies (Shier et al. 1976; Kitao and Hattori 1977; Yamaguchi et al. 1979; Lin et al. 1981; Liautard 1985; Lin and Lin 1985; Shier 1985).

Although they tolerate high coupling frequencies without loss of binding activity and show prolonged retention at the site of injection into tissue, their lack of general high cell type specificity, sufficient for cell separation in certain cases (Sharon 1983; Lis and Sharon 1986; Gabius et al. 1988c), and their often inherent toxicity have limited their clinical applicability. Specificity could be enhanced by custom-made carriers, refined to exhibit the indispensable degree of target selectivity without intrinsic toxicity. Since neoglycoproteins have already been described above as suppositionally recommendable tools in lectin-based histopathological diagnosis, the presence of lectins in tumors may contribute to improvement of drug delivery by achieving selective lectin-mediated binding and, moreover, uptake of therapeutically active glycoproteins.

The carrier potential of glycoproteins in this system of specific protein-carbohydrate interaction, where the carbohydrate part can be synthetically adapted to meet the clinical requirements, may thus be a promising perspective in the field of drug targeting. Whereas the carbohydrate moieties on the carrier render the conjugate accessible to lectin-mediated binding and uptake, the cotransported drug will be released after intracellular proteolysis of the carrier protein. Concerning receptors, screening for the presence of membrane lectins in tumor cells is facilitated by the availability of a large number of both tumor cell lines and labeled, suitably modified neoglycoproteins (Monsigny et al. 1983, 1984, 1988; Kieda and Monsigny 1986; Gabius et al. 1988c). The latter mentioned tools can also be effectively used as a comparative analysis of the pattern of expression of membrane lectins in normal cells as diverse as myeloid cells or spermatozoa (Monsigny et al. 1979, 1988; Sinowatz et al. 1988) or in various functional states of any cell type. For example, fluorescent neoglycoproteins help to quantify changes in membrane lectin expression by comparative analysis of peripheral blood lymphocytes before and after stimulation (Table 5.6). The degree of stimulation is constantly monitored by cell cycle analysis, using simultaneous measurement of DNA/RNA-content, illustrated in Figure 5.4. This technique, using the metachromatic fluorochrome acridine orange, has been described by Andreeff et al. (1980). As alternative label for the detection of membrane lectins, radioiodination of neoglycoproteins can be considered, providing highly sensitive probes. Their usefulness is exemplified in the case of Ehrlich ascites tumor cells (Table 5.7).

Following screening for the presence of membrane lectins, the thereby elucidated specific binding and uptake signal can be added to carrier-drug conjugates in the form of various sugars. The effectiveness of the neoglycoprotein-drug conjugates can be measured by their tumoricidal potency. In the case of human embryonic carcinoma cells, simple addition of mono- or disaccharides to an etoposide (VP-16)-protein conjugate, synthesized by coupling the activated hemisuccinate derivative of the drug to the carrier protein (Fig. 5.5), confers a greater than 10-fold increase in growth-inhibitory capacity to the conjugate relative to the non-glycosylated drug-carrier complex (Fig. 5.6). Similar results, e.g., obtained with colon carcinoma cells and 5-fluorodeoxyuridine (FuDr)-neoglycoprotein conjugates (Table 5.8) or virally transformed fibroblasts and methotrexate as cotransported chemotherapeutical (Fig. 5.7), underscore the principal feasibility and potential of lectin-directed drug targeting for tumor cells (Roche et al. 1983; Schneider et al. 1983; Monsigny et al. 1984; Gabius et al. 1986b, 1987c, d; Gabius and Vehmeyer 1987; Gabius 1988). To reduce the extent or prevent conjugates from reaching undesired target cells, the specificity of tumor lectins should be clearly determined. As far as the present state of knowledge indicates, certain endogenous lectins in tumors may differ at least quantitatively in their level of expression compared to non-malignant cells (Gabius 1987b). Consistent construction of carriers with custom-made sugar part and, consequently, improved deployment capability for the drug could then take advantage of any detectable difference in the sugar-binding capability of tumor cells in comparison to normal cells.

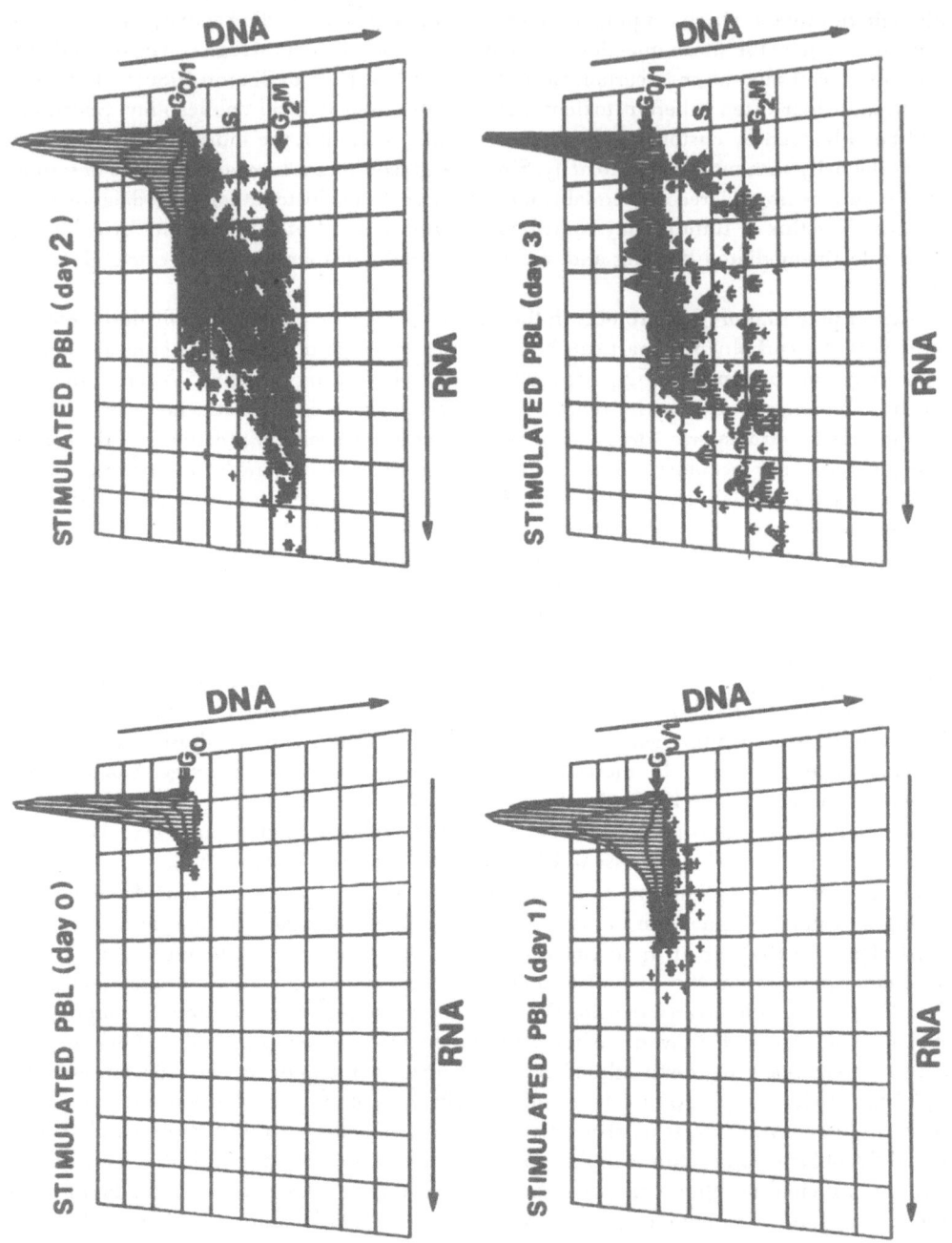

Fig. 5.4 Stimulation of human peripheral blood lymphocytes (PBL) by interleukin 2 and OKT 3 mono-clonal antibody, monitored by simultaneous measurement of cellular DNA (F_{530}) and RNA ($F_{>600}$). A maximal number of cells in cell cycle phases S and (G_2M) is present after 2 days

Neoglycoprotein	Peripheral lymphocytes	
	Unstimulated	Stimulated
lac(diazo)-FBSA	14	5
lac(red. am.)-FBSA	12	2
mel-FBSA	11	11
ASF	31	28
ALF	7	19
α-L-fuc-FBSA	7	4
α-galNAc-FBSA	12	5
mal-FBSA	22	15
α-D-man-FBSA	7	4

Table 5.6 Percentage of (FITC-neoglyco-protein)-conjugate-binding cells
lac: lactose (β-galactoside), diazo: diazotation, red. am.: reductive amination, mel: melibiose (α-galactoside), ASF: asialofetuin, ALF: asialolactoferrin, fuc: fucose (α-fucoside), galNAc: N-acetylgalactosamine, mal: maltose (α-glucoside), man: mannose (α-mannoside)
Background (binding of non-glycosylated FBSA) was substracted to give the values shown.

Coupled sugar	Binding [%]
lac	4.16 ± 0.29
β-D-Gal	1.11 ± 0.18
mel	6.91 ± 1.13
α-D-gal	2.44 ± 0.80
glcNAc	5.58 ± 0.82
galNAc	2.44 ± 0.71
L-fuc	2.00 ± 0.36
D-man	6.07 ± 0.64
L-rham	4.02 ± 0.31
cel	3.39 ± 0.18
β-D-glc	1.49 ± 0.15
man-6-P	7.08 ± 0.99
gal-6-P	2.13 ± 0.24
Sialic acid	5.70 ± 1.00
Glucuronic acid	8.31 ± 1.12
D-xyl	9.59 ± 0.94

Table 5.7 In vitro binding of ^{125}I-neoglycoproteins by Ehrlich ascites tumor cells
Each value represents mean binding $\pm$ SD for 4 independent assays, expressed as percentage of total radioactivity, added to the cells.
lac: lactose, gal: galactose, mel: melibiose, glcNAc: N-acetylglucosamine, galNAc: N-acetylgalactosamine, fuc: fucose, man: mannose, rham: rhamnose, cel: cellobiose, glc: glucose, P: phosphate, xyl: xylose

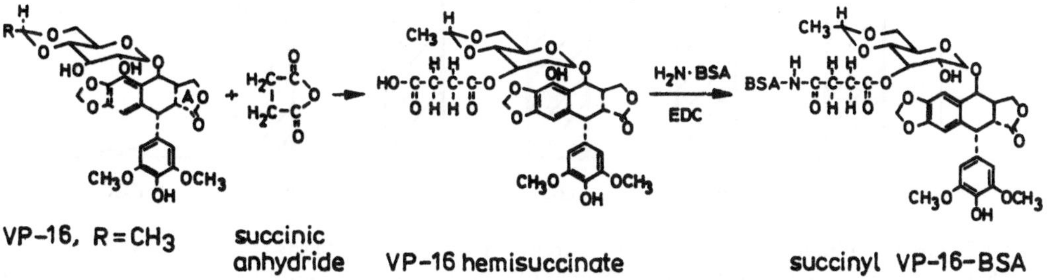

VP–16, R = CH₃   succinic anhydride   VP–16 hemisuccinate   succinyl VP–16–BSA

Fig. 5.5 Scheme for the mechanism of activation of etoposide (VP-16) by succinic anhydride and conjugation of the resulting VP-16 hemisuccinate to bovine serum albumin (BSA) in the presence of 1-ethyl-3 (3-dimethyl-aminopropyl) carbodiimide hydrochloride (EDC), resulting in succinyl VP-16-BSA

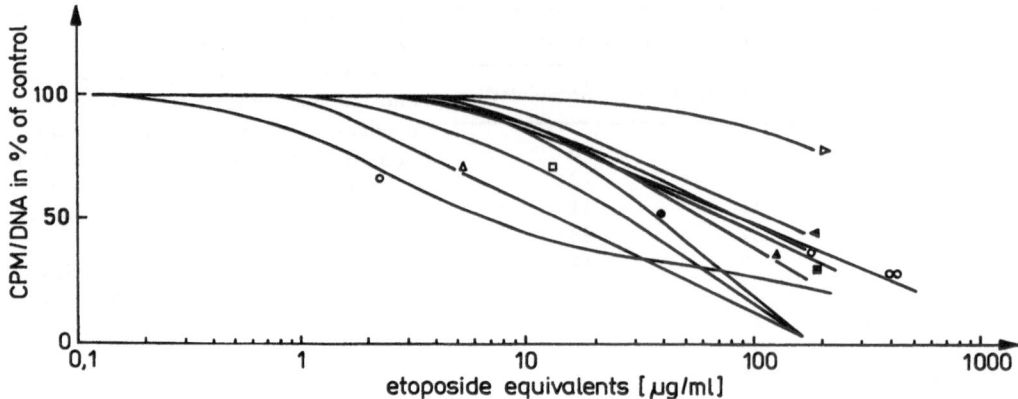

Fig. 5.6 Growth inhibition of H 12.1 cells, determined by measuring ^{3}H-thymidine incorporation. [Free etoposide (o), BSA-bound etoposide (∞) and neoglycoprotein-bound etoposide: melibiosylated BSA (△), maltosylated BSA (□), lactosylated BSA (●), fucosylated BSA (▲), cellobiosylated BSA (■), mannosylated BSA (o), xylosylated BSA (◀), and BSA, carrying mannose-6-phosphate (▷)] Each curve drawn from 6–8 points over the concentration range, showing SD <8%

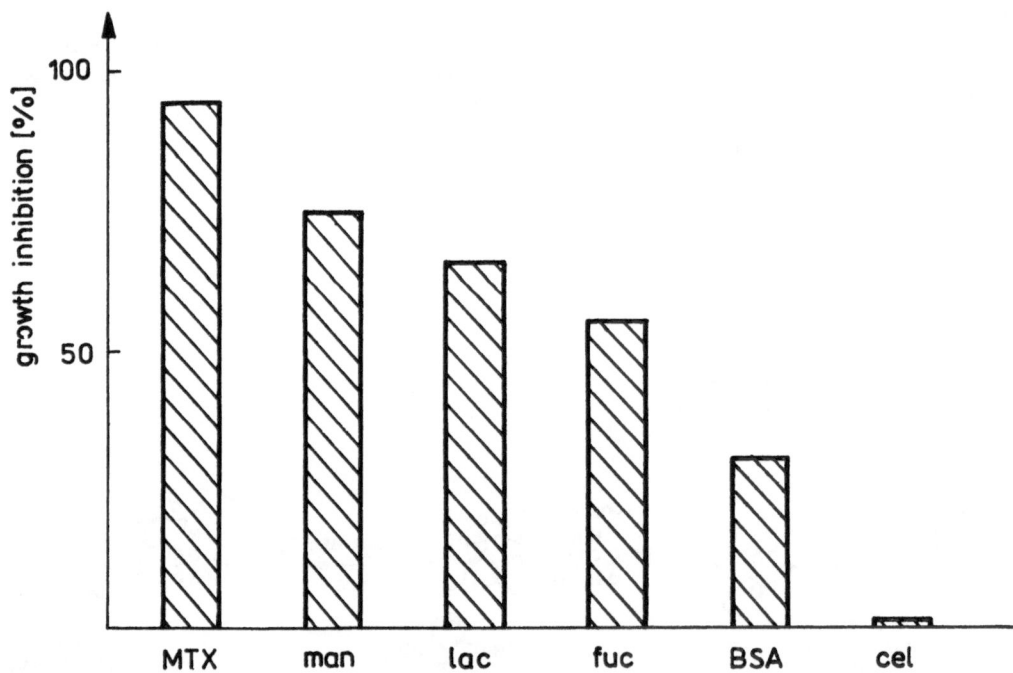

Fig. 5.7 Growth inhibition of tumor cells of virally transformed fibroblasts (clone 5–20#20) by 1 μM methotrexate (MTX) and 1 μM bound MTX in MTX-neoglycoprotein conjugates. MTX-neoglycoprotein conjugates were prepared with bovine serum albumin (BSA) and glycosylated BSA: mannosylated BSA (man), lactosylated BSA (lac), fucosylated BSA (fuc) and cellobiosylated BSA (cel)

Neoglycoprotein	SW 480 [%]	SW 620 [%]
FuDr	62	90
BSA	4	26
α-glc-BSA	30	63
α-man-BSA	15	6
α-L-fuc-BSA	20	30
β-lac-BSA	19	20
α-gal-BSA	62	57
β-glcNAc-BSA	40	51

Table 5.8 Growth inhibition of tumor cells by FuDr and FuDr-neoglycoprotein conjugates at a FuDr concentration of 10^{-6} M

FuDr: 5-fluorodeoxyuridine, BSA: bovine serum albumin, glc: glucose, man: mannose, fuc: fucose, lac: lactose, gal: galactose, glcNAc: N-acetylglucosamine

Straightforwardly, such modifications in the target control part of neoglycoproteins can be tested either with cells by trial-and-error experiments or with endogenous lectins, purified by affinity chromatography, as performed for human colon carcinoma cells on supports with immobilized carbohydrates (Fig. 5.8). In this respect, it is noteworthy that the yields of endogenous tumor lectins can be drastically dependent on the type of immobilized affinity ligand (Gabius and Engelhardt 1987). Careful selection among possible choices can pay off well in terms of increased quantitative yields, enabling specificity studies of tumor lectins with less starting material, as comprehensively compiled for six types of affinity ligand in Table 5.9.

Since successful performance of initial drug targeting experiments to tumor cells in vitro will not necessarily translate into effective drug delivery and tumor eradication in vivo, it is just and appropriate to add a cautious note on the future of this branch of tumor lectinology. However, identifying and defining significant problems will now allow a systematic series of experiments to try and design satisfying solutions to the pending problems.

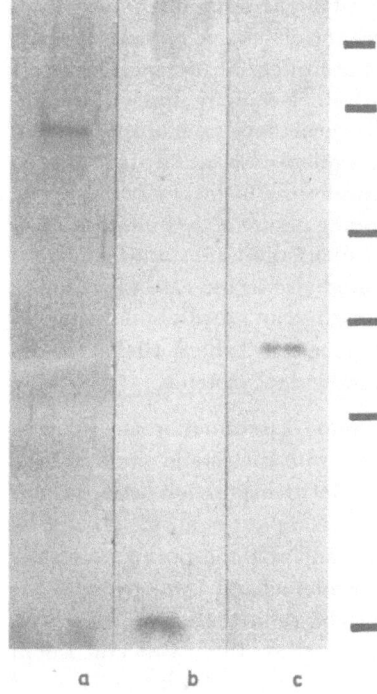

a   b   c

Fig. 5.8 Polyacrylamide gel electrophoresis in the presence of 0.1 % sodium dodecyl sulfate of endogenous lectins from tumors, grown in nude mice, of the human colon carcinoma cell line SW 620. Elution with EDTA from immobilized melibiose (a), elution with specific sugar from immobilized lactose (b) and immobilized maltose (c). Standards for molecular weight designations are indicated by bars: phosphorylase b (MW 97,000), bovine serum albumin (MW 66,000), ovalbumin (MW 44,000), glyceraldehyde-3-phosphate dehydrogenase (MW 36,000), carbonic anhydrase (MW 29,000) and lysozyme (MW 14,300)

Table 5.9 Effect of the type of immobilized ligand on the yield of Ca^{2+}-dependent mannose-binding proteins[1]

Tissue	Type of ligand					
	BSA_4^2	BSA_{18}^3	RNase B	Mannan	Invertase	D-Mannose
Human liver	30	46	63	86	68	89
Rat fibroadenoma	12	17	19	28	31	202
Roe liver	34	82	450	680	585	1250

[1]: Yield is given in μg per 5 g liver or tumor.
[2]: Mannosylated bovine serum albumin with 4 sugar residues.
[3]: Mannosylated bovine serum albumin with 18 sugar residues.

5.4   Can Tumor Lectins Lead to the Discovery of Pieces to be Integrated into the Challenging Puzzle of Understanding the Mechanisms of Tumor Spread?

Development of initial diagnostic and therapeutic procedures, focusing on endogenous tumor lectins, are based on the idea that their profile may, in biochemical terms, reflect aspects of the malignant phenotype. In contrast to the monoclonal antibody approach, with pragmatic selection for the capability to uncover immunologically recognizable differences in tumor cells compared to various types of normal cells, the concept for tumor lectins hypothetically ascribes definite functional relevance to them, that subsequently needs to be experimentally confirmed or dismissed. Since within tumor biology the factors influencing growth and metastatic behavior have so far largely remained enigmatic, it seems promising to test this assumption thoroughly.

Relating these aspects of tumor biology to the described application of neoglycoproteins, refinement of their fine structural features must still achieve more precise identification of certain lectins within a group of proteins with rather similar carbohydrate specificity. Indeed, the approach of classifications with the currently used glycohistochemical tools is not able to discern accurately between such related sugar receptors. Therefore, adequate description of the pattern of lectins biochemical analysis. This type of analysis corroborates the histochemically drawn conclusions on profile changes in lectin expression. The changing nature of the composition of endogenous sugar receptors within tumor development or differentiation, substantiated for rat mammary tumors (Gabius et al. 1986c), is further documented also in the case of a human model of non-malignant origin with limited resemblance to malignant growth both histochemically and biochemically (Gabius et al. 1987e). It can clearly be seen in Table 5.10 that the developmental changes take place on the level of expression of individual proteins.

To pinpoint cellular properties that may partly account for tumor dissemination and growth, biochemical analyses of various murine tumor model systems with metastatic variants have positively addressed the question whether changes can occur in lectin expression between cells of similar genetic background, but divergent metastatic potency (Gabius et al. 1986b, 1987f; Raz et al. 1987). Inhibition of anchorage-independent growth and of blood-borne metastasis for B16 melanoma cells and UV-2237 fibrosarcoma cells by a monoclonal antibody to an endogenous lectin imply involvement of a galactoside-specific lectin in mediation of cell-to-cell and/or cell-to-substratum adhesion and growth control (Lotan et al. 1985; Raz and Lotan 1987). Regulation of cell proliferation by soluble endogenous lectins has also been discussed

Table 5.10 Sugar-binding proteins of human placenta[1]

	EDTA						Sugar					
	lac	ASF	mel	man	fuc	mal	lac	ASF	mel	man	fuc	mal
Human placenta (full-term)	29	29	56, 29	29	180, 70, 56	–	14	–	52	–	52 28	52, 31, 28
	29	29	29	29	180 29	29	14	–	52	–	–	34 31
Trophoblastic layer (first term)	72, 67, 29, 14	–		97, 92, 60	97, 92, 60, 26.5 25	74, 65, 60	14	–	52, 29	–	52	74, 65, 52, 34 31
	97, 94, 29, 90	–		–	–	–	14	–	–	–	–	34, 31

The extraction and elution conditions permit a dividing of the pattern into salt extractable (first line) or detergent extractable (second line), Ca^{2+}-dependent (elution by EDTA) or Ca^{2+}-independent (elution by specific sugar) sugar-binding proteins.

[1]: Apparent molecular weight is given in thousands.

lac: lactose, ASF: asialofetuin, mel: melibiose, man: mannan, fuc: fucose, mal: maltose

Table 5.11 Carbohydrate-binding proteins of human liver metastases[1]

	EDTA					Sugar				
	lac	ASF	mel	man	fuc	lac	ASF	mel	man	fuc
Small cell lung	21, 22, 28 42, 58, 60 95, 100	28	28, 54 56	28, 42	31, 72	14, 54	54	54	–	–
carcinoma		28	–	–	28	14	–	54	–	–
Colon	12, 15, 16, 28 34, 66	28	28, 66	28	28, 66, 80	14, 15, 54, 66	26, 54	54	–	66
carcinoma	66	28	66	28	66	14, 21, 66	–	54, 66	–	66

The extraction and elution conditions permit a dividing of the pattern into salt extractable (first line) or detergent extractable (second line), Ca^{2+}-dependent (elution by EDTA) or Ca^{2+}-independent (elution by specific sugar) lectins.

[1]: Apparent molecular weight is given in thousands.

lac: lactose, ASF: asialofetuin, mel: melibiose, man: mannan, fuc: fucose

with respect to tumor biology by Caron et al. (1988). Moving from murine tumor model systems to the clinical situation of human metastases, profile differences as possible trail of functional relevance have been determined for various types of metastases, as outlined in Table 5.11 for two metastatic lesions to liver, originating from different primary sites. The detailed compila-

Table 5.12 Sugar-binding proteins of two murine macrophage-like cell lines[1]

	Lactose	L-Fucose	N-Acetyl-D-glucosamine	N-Acetyl-D-galactosamine	Maltose	Mannan	Sialic acid
P388D$_1$ (culture)	31[2], 14	22	–	–	–	22	21.5
J774A.1 (culture)	14	66, 22	–	66	–	66, 22	–
P388D$_1$ (tumors in nude mice)	31	92, 88, 59, 57	29.5	–	–	80, 66	64, 30
P388D$_1$ (re-adapted to in vitro growth)	14	22	–	–	34	32, 22	–

[1]: Relative molecular masses are given in thousands. [2]: very weak band

tion of lectin profiles for various types of human metastases to lung and liver has led to the conclusion that metastatic lesions to the same target organ, originating from different types of primary tumor, resemble each other more in this respect than metastatic lesions to different target organs of the same type of primary tumor (Gabius and Engelhardt 1988). Nevertheless, this result cannot necessarily be interpreted as indication of lectin involvement in establishing the organotropy of metastasis.

Besides dependence of homing and growth at secondary sites of malignant cells on tumor cell characteristics, the microenvironment of the target organ can itself exert a strong influence on quantitative and qualitative phenotypic aspects of the malignant cells (Fidler and Balch 1987; Nicolson 1987). Modulation of lectin expression numbers among such flexible cellular characteristics, as shown for a cell line grown under different conditions (Table 5.12). These results emphasize that no prompt conclusion can be expected on the relevance of endogenous lectins in metastasis, although they appear to suggest that distinct protein-carbohydrate interactions may contribute to the overall success of a tumor cell's ability to perform all required steps within the metastatic cascade. However, due to the complexity of the involved individual interactions with normal host cells, tissues and cellular and physical factors and the heterogeneity of tumor cells (Sugarbaker 1979; Schirrmacher 1985; Weiss 1985), results of tumor lectinology need to be integrated into the complex puzzle among many other observations. Its solution, if only partially, holds out the conceivable prospect of setting up rational schemes to combat tumor growth and spread.

5.5 Are There Perspectives in Tumor Lectinology?

Understanding the specific molecular interactions that govern the social behavior of cells is of central scientific and clinical relevance. The reviewed lines of research in tumor lectinology appear to warrant a critical evaluation of the potential merits, aimed at translating any progress in basic science into clinical benefit. Only further tenacious efforts in this field will prove whether the optimistic attitude that a clinically favorable role of endogenous tumor lectins may only just have begun to unfold is justified. Thereby the field of tumor lectinology will unquestionably add another exciting chapter to the eventful history of lectinology, that started 100 years ago.

Acknowledgement.
I greatly appreciate the excellent secretarial assistance of U. Rust and the financial support of the Dr.-Mildred-Scheel-Stiftung für Krebsforschung. I am also much obliged to my enthusiastic co-workers and to the colleagues of various disciplines, with whom we have had the privilege of cooperating.

5.6 References

Allison RT (1986) Lectins in diagnostic histopathology: a review. Med Lab Sci 43:369–376

Andreeff M, Darzynkiewicz Z, Sharpless T, Clarkson B, Melamed MR (1980) Discrimination of human leukemia subtypes by the flow cytometric analysis of cellular DNA and RNA. Blood 56:282–293

Aplin JD, Wriston JC (1981) Preparation, properties, and applications of carbohydrate conjugates of proteins and lipids. CRC Crit Rev Biochem 10:259–306

Bardosi A, Dimitri T, Gabius HJ (1988a) Endogenous carbohydratebinding proteins in neuro-oncology. In: Gabius HJ, Nagel GA (eds) Lectins and glycoconjugates in oncology. Springer, Berlin Heidelberg New York Tokyo, pp 143–152

Bardosi A, Dimitri T, Gabius HJ (1988b) Endogenous carbohydratebinding proteins in oligodendrogliomas. A histochemical study. Acta Neuropathol, (in press)

Barondes SH (1986) Vertebrate lectins: properties and functions. In: Liener IE, Sharon N, Goldstein IJ (eds) The lectins: properties, functions and applications in biology and medicine. Academic Press, Orlando, pp 437–465

Berger EG, Buddecke E, Kamerling JP, Kobata A, Paulson JC, Vliegenthart JFG (1982) Structure, biosynthesis and functions of glycoprotein glycans. Experientia 38:1129–1158

Brisson JR, Carver JP (1983) The relation of three-dimensional structure to biosynthesis in the N-linked oligosaccharides. Can J Biochem Cell Biol 61:1067–1078

Cairns J (1985) The treatment of diseases and the war against cancer. Sci Am 253 (5):31–39

Caron M, Joubert R, Bladier D (1988) Soluble lectins and endothelial cell growth factors. In: Gabius HJ, Nagel GA (eds) Lectins and glycoconjugates in oncology. Springer, Berlin Heidelberg New York Tokyo, pp 179–186

Caselitz J (1987) Lectins and blood group substances as "tumor markers". Curr Top Pathol 77:245–278

Coggi G, Dell'Orto P, Bonoldi E, Doi P, Viale G (1983) Lectins in diagnostic pathology. In: Bøg-Hansen TC, Spengler GA (eds) Lectins: biology, biochemistry and clinical biochemistry, vol 3. De Gruyter, Berlin (W), pp 87–103

Cohen MM, Diamond JM (1986) Are we losing the war on cancer? Nature (London) 323:488–489

Cook GMW (1986) Cell surface carbohydrates: molecules in search of a function? J Cell Sci Suppl 4:45–70

Cooper HS (1984) Lectins as probes in histochemistry and immunohistochemistry: the peanut (Arachis hypogaea) lectin. Hum Pathol 15:904–906

Damjanov I (1987) Lectin cytochemistry and histochemistry. Lab Invest 57:5–20

Debbage PL, Lange W, Hellmann T, Gabius HJ (1988) Receptors for sulfated polysaccharides in human placenta. J Histochem Cytochem (in press)

Feizi T (1985) Demonstration by monoclonal antibodies that carbohydrate structures of glycoproteins and glycolipids are onco-developmental antigens. Nature (London) 314:53–57

Feizi T, Childs RA (1987) Carbohydrates as antigenic determinants of glycoproteins. Biochem J 245:1–11

Fidler IJ, Balch CM (1987) The biology of cancer metastasis and implications for therapy. Curr Probl Surg 24:132–209 :

Freeman AI, Mayhew E (1986) Targeted drug delivery. Cancer 58:573–583

Gabius HJ (1987a) Vertebrate lectins and their possible role in fertilization, development and tumor biology. IN VIVO 1:75–84

Gabius HJ (1987b) Endogenous lectins in tumors and the immune system. Cancer Invest 5:39–46

Gabius HJ (1988) Targeting of neoglycoprotein-drug conjugates to human tumor cells via endogenous lectins. Ann NY Acad Sci (in press)

Gabius HJ, Engelhardt R (1987) Interaction of mannose-binding proteins with different types of immobilized affinity ligand. J Chromat 391:452–456

Gabius HJ, Vehmeyer K (1987) Membrane lectins in human malignant melanoma. Naturwissenschaften 74:37–38

Gabius HJ, Engelhardt R (1988) Sugar receptors of different types in human metastases to lung and liver. Tumor Biol 9:21–36

Gabius HJ, Brehler R, Schauer A, Cramer F (1986a) Localization of endogenous lectins in normal human breast, benign breast lesions and mammary carcinomas. Virch Arch [Cell Pathol] 52:107–115

Gabius HJ, Vehmeyer K, Engelhardt R, Nagel GA, Cramer F (1986b) Carbohydrate-binding proteins of tumor lines with different growth properties. II. Changes in their pattern in clones of transformed rat fibroblasts of differing metastatic potential. Cell Tiss Res 246:515–521

Gabius HJ, Engelhardt R, Rehm S, Deerberg F, Cramer F (1986c) Differences in the pattern of endogenous lectins from spontaneous rat mammary tumors. Tumour Biol 7:71–81

Gabius HJ, Gabius S, Vehmeyer K, Schauer A, Nagel GA (1987a) Endogene Tumorlektine: Neue Tumormarker und Zielpunkte für Therapie? Onkologie 10:184–185

Gabius HJ, Engelhardt R, Hellmann KP, Hellmann T, Ochsenfarth A (1987b) Preparation of neoglycoprotein-enzyme conjugate using a heterobifunctional reagent and its use in solid-phase assays and histochemistry. Anal Biochem 165:349–355

Gabius HJ, Bokemeyer C, Hellmann T, Schmoll HJ (1987c) Targeting of neoglycoprotein-drug conjugates to cultured human embryonic carcinoma cells. J Cancer Res Clin Oncol 113:126–130

Gabius HJ, Engelhardt R, Hellmann T, Midoux P, Monsigny M, Nagel GA, Vehmeyer K (1987d) Characterization of membrane lectins in human colon carcinoma cells by flow cytofluorometry, drug targeting and affinity chromatography. Anticancer Res 7:109–112

Gabius HJ, Debbage PL, Engelhardt R, Osmers R, Lange W (1987e) Identification of endogenous sugar-binding proteins (lectins) in human placenta by histochemical localization and biochemical characterization. Eur J Cell Biol 44:265–272

Gabius HJ, Bandlow G, Schirrmacher V, Nagel GA, Vehmeyer K (1987f) Differential expression of endogenous sugar-binding proteins (lectins) in murine tumor model systems with metastatic capacity. Int J Cancer 39:643–648

Gabius HJ, Bodanowitz S, Schauer A (1988a) Endogenous sugar-binding proteins in human breast tissue and benign and malignant breast lesions. Cancer 61:1125–1131

Gabius HJ, Kohnke B, Hellmann T, Dimitri T, Bardosi A (1988b) Comparative histochemical and biochemical analysis of endogenous receptors for glycoproteins in human and pig peripheral nerve. J Neurochem (in press)

Gabius HJ, Vehmeyer K, Gabius S, Nagel GA (1988c) Clinical application of various plant and endogenous lectins to leukemia. Blut (in press)

Hakomori SI (1981) Glycosphingolipids in cellular interaction, differentiation and oncogenesis. Annu Rev Biochem 50:733–764

Hascall VC (ed) (1986) Functions of the proteoglycans. Ciba Found Symp 124, Wiley-Interscience, Chichester

Höök M, Kjellen L, Johansson S (1984) Cell-surface glycosaminoglycans. Annu Rev Biochem 53:847–869

Homans SW, Dwek RA, Rademacher TW (1987) Solution conformations of N-linked oligosaccharides. Biochemistry 26:6571–6578

Iozzo R (1985) Proteoglycans: structure, function and role in neoplasia. Lab Invest 53:373–396

Kataoka M, Tavassoli M (1985) Identification of lectin-like substances recognizing galactosyl residues of glycoconjugates on the plasma membrane of marrow sinus endothelium. Blood 65:1163–1171

Kayser K and Gabius HJ (1988) Histomorphological characterization of carbohydrate-binding proteins in human lung cancer. In: Gabius HJ, Nagel GA (eds) Lectins and glycoconjugates in oncology. Springer, Berlin Heidelberg New York Tokyo, pp 131–142

104

Kieda C, Monsigny M (1986) Involvement of membrane sugar receptors and membrane glycoconjugates in the adhesion of 3LL cell subpopulations to cultured pulmonary cells. Invasion Metastasis 6:347–366

Kitao T, Hattori K (1977) Concanavalin A as carrier of daunomycin. Nature (London) 265:81–82

Kolb-Bachofen V, Schlepper-Schäfer J, Vogell W (1982) Electron microscopic evidence for an asialoglycoprotein receptor on Kupffer cells: localization of lectin-mediated endocytosis. Cell 29:859–866

Liautard JP, Vidal M, Philippot JR (1985) Controlled binding of liposomes to cultured cells by means of lectins. Cell Biol Internat Rep 9:1123–1137

Lin JY, Lin LL (1985) Antitumor lectin-trypsin inhibitor conjugate. J Natl Cancer Inst 74:1031–1036

Lin JY, Li JS, Tung TC (1981) Lectin derivatives of methotrexate and chlorambucil as chemotherapeutic agents. J Natl Cancer Inst 66:523–528

Lis H, Sharon N (1986) Lectins as molecules and as tools. Annu Rev Biochem 55:35–67

Lotan R, Lotan D, Raz A (1985) Inhibition of tumor cell colony formation in culture by a monoclonal antibody to endogenous lectins. Cancer Res 45:4349–4353

Miettinen M, Holthöfer H, Lehto VP, Miettinen A, Virtanen I: Ulex europaeus I lectin as a marker for tumors derived from endothelial cells. Am J Clin Pathol 79:32–36

Monsigny M, Kieda C, Roche AC (1979) Membrane lectins. Biol Cell 36:289–300

Monsigny M, Kieda C, Roche AC (1983) Membrane glycoproteins, glycolipids and membrane lectins as recognition signals in normal and malignant cells. Biol Cell 47:95–110

Monsigny M, Roche AC, Kieda C (1984) Uptake of neoglycoproteins via membrane lectin(s) of L1210 cells evidenced by quantitative flow cytofluorometry and drug targeting. Biol Cell 51:187–196

Monsigny M, Roche AC, Midoux P, Kieda C, Mayer R (1988) Endogenous lectins of myeloid and tumor cells: characterization and clinical implications. In: Gabius HJ, Nagel GA (eds) Lectins and glycoconjugates in oncology. Springer, Berlin Heidelberg. New York Tokyo, pp 25–47

Montreuil J (1986) Structure and conformation of glycoprotein glycans. In: Olden K, Parent JB (eds) Vertebrate lectins. Van Nostrand Reinhold Co., New York, pp 1–26

Muramatsu T (1984) Cell surface glycoproteins as markers in monitoring in vitro differentiation of embryonal carcinoma cells. Cell Differentiation 15:101–108

Nicolson GL (1987) Tumor cell instability, diversification and progression to the metastatic phenotype: from oncogene to oncofetal expression. Cancer Res 47:1473–1487

Poznansky MJ, Cleland LG (1980) Biological macromolecules as carriers of drugs and enzymes. In: Juliano RL (ed) Drug delivery systems. Oxford Univ, New York, pp 253–315

Poste G (1983) Drug targeting in cancer therapy. In: Gregoriadis G, Poste G, Senior J, Trouet A (eds) Receptor-mediated targeting of drugs. Plenum, New York London, pp 427–474

Raz A, Lotan R (1987) Endogenous galactoside-binding lectins: a new class of functional tumor cell surface molecules related to metastasis. Cancer Metastasis Rev 6:433–452

Raz A, Meromsky L, Zvibel I, Lotan R (1987) Transformation-related changes in the expression of endogenous cell lectins. Int J Cancer 39:353–360

Roche AC, Barzilay M, Modoux P, Junqua S, Sharon N, Monsigny M (1983) Sugar-specific endocytosis of glycoproteins by Lewis lung carcinoma cells. J Cell Biochem 22:131–140

Sairam MR (1985) Protein glycosylation and receptor-ligand interaction. In: Conn RA (ed) Receptors, vol 2. Academic Press, Orlando, pp 307–340

Schirrmacher V (1985) Cancer metastasis: experimental approaches, theoretical concepts and impacts for treatment strategies. Adv Cancer Res 43:1–73

Schneider YJ, Abarca J, Aboud-Pirak E, Baurain R, Ceulemans F, Deprez-De Campaneere D, Lesur B, Masquelier M, Otte-Slachmuylder C, Rolin-van Swieten D, Trouet A (1983) Drug targeting in human cancer chemotherapy. In: Gregoriadis G, Poste G, Senior J, Trouet A (eds) Receptor-mediated targeting of drugs. Plenum, New York London, pp 1–25

Sharon N (1983) Lectin receptors as lymphocyte surface markers. Adv Immunol 34:213–298

Shier WT (1985) Lectins as carriers: preparation and purification of a concanavalin A-trypsin conjugate. Meth Enzymol 112:248–258

Shier WT, Trotter JT, Astudillo DT (1976) Effect of localization of L-asparaginase as the concanavalin A-conjugate on anti-tumor activity. Int J Cancer 18:672–678

Sinowatz F, Gabius HJ, Amselgruber W (1988) Surface sugar-binding components of bovine spermatozoa as evidenced by fluorescent neoglycoproteins. Histochemistry 88:395–399

Stowell CP, Lee YC (1980) Neoglycoproteins. The preparation and application of synthetic glycoproteins. Adv Carbohydr Chem Biochem 37:225–281

Sugarbaker EV (1979) Cancer metastasis: a product of tumor-host interactions. Curr Probl Cancer 3:1–59

Weiss L (1985) Principles of metastasis. Academic Press, Orlando

Yamaguchi T, Kato R, Beppu M, Terao T, Inoue Y, Ikawa Y, Osawa T (1979) Preparation of concanavalin A-ricin A-chain conjugate and its biologic activity against various cultured cells. J Natl Cancer Inst 62:1387–1395

Index

107

Volume 3 – 1990

Volume 4 – 1991